Earth and Environmental Sciences Library

Earth and Environmental Sciences Library (EESL) is a multidisciplinary book series focusing on innovative approaches and solid reviews to strengthen the role of the Earth and Environmental Sciences communities, while also providing sound guidance for stakeholders, decision-makers, policymakers, international organizations, and NGOs.

Topics of interest include oceanography, the marine environment, atmospheric sciences, hydrology and soil sciences, geophysics and geology, agriculture, environmental pollution, remote sensing, climate change, water resources, and natural resources management. In pursuit of these topics, the Earth Sciences and Environmental Sciences communities are invited to share their knowledge and expertise in the form of edited books, monographs, and conference proceedings.

Fortunato Garza-Ocañas
Editor

Sustainable Management of Natural Resources

Diversity, Ecology, Taxonomy and Sociology

Springer

Editor
Fortunato Garza-Ocañas
Faculty of Forestry Sciences, Department
of Sylviculture
Universidad Autónoma de Nuevo León
Nuevo León, Mexico

ISSN 2730-6674 ISSN 2730-6682 (electronic)
Earth and Environmental Sciences Library
ISBN 978-3-031-33396-5 ISBN 978-3-031-33394-1 (eBook)
https://doi.org/10.1007/978-3-031-33394-1

This Springer imprint is published by the registered company Springer Nature Switzerland AG
The registered company address is: Gewerbestrasse 11, 6330 Cham, Switzerland

Survival of species is a need for humans.

This book is dedicated to the memory of Professor Rolando Guerra González who was a lecturer at the Faculty of Forestry Sciences of the Universidad Autónoma de Nuevo León. He wholeheartedly supported and contributed to the development of many students who became graduated professors from several national and international universities. As a lecturer, he always promoted that a critical and positive mind is a needed tool for every student to contribute to the advancement of science in every subject.
The authors

Foreword

The Universidad Autónoma de Nuevo León (UANL) trains professionals in many areas of science, this is in accordance with the UANL's Vision 2030 that it is oriented to promote human values to generate education and research of quality for the benefit of society.

Our institution has high quality standards at national and international levels and brings together the talents and work of teachers and researchers committed with education for the development of students.

The University encourages continuous education training for teaching-research staff as a fundamental tool to support students as well as to carry out scientific research that impacts the society daily. In relation to forest ecosystems and resources, it is important to produce research and technology allowing a greater efficiency for their sustainable management and conservation.

This is an alternative to generate opportunities for people living in rural and forest areas. The new generations of forestry professionals are well trained in the sustainable use and management of forest resources.

The UANL has the pleasure of providing the lecturers with this book that contains relevant and updated information regarding the management, diversity, ecology, taxonomy, and sociology of forest and people. The chapters of this book were written by researchers and professors of the academic body of Natural Resources Management and Sustainability of the Faculty of Forest Sciences who, in collaboration with authors from several institutions, worked together and shared their valuable experience and knowledge on natural resources and promoted their sustainable management for the benefit of students and the society.

Rector Universidad Autónoma de Nuevo León, Nuevo Leon, Mexico

Santos Guzmán López M.D., PhD.

Introduction

Celebrating 40 years of the sowing of the seed, of the Faculty of Forestry Sciences with the edition of a book, I consider it a great success. The topics addressed demonstrate the variety of interests, studies, and research of this faculty team, as well as the wide range of opportunities for students. I can only thank you for the invitation to write the introduction to this book.

For those who have known the faculty since its early days, and among whom I count, it is a great satisfaction to see it formed as a center of excellence in the sciences and in the management of forest resources. This Faculty had its origin in an idea of the Rector, Dr. Alfredo Piñeyro, in 1980. Or rather three ideas: to promote faculties of applied sciences, to initiate the training of well-qualified teachers before receiving the first students, and to decentralize the University to Linares.

Other Forestry Faculties in Mexico covered the management and forestry of temperate forests, so the first programs of the curriculum were designed for this region of the country: with its arid and semi-arid zones, and its temperate forests with enormous environmental value, but with less commercial importance. The reason was a vision to improve the management of these resources and lands in the properties, ranches, ejidos, municipalities, and states of the region. This is how the faculty began, with great enthusiasm: something traditional in some senses, but with international approaches, and with a balance of women and men students not so conventional. In the first few years, the studies of teachers and fellows were focused on acquiring understandings of forest ecosystems and resources, in the broad sense, and their dynamics.

But the forest world has changed and continues to change. In the 1980s, there was strong concern from NGOs and the public about the rates of deforestation, and biodiversity losses, associated with forestry and agricultural practices. Governments and international organizations were beginning to develop more appropriate agreements and policies and (little by little) carry them out. Concerns about climate change were still limited to think tanks and a few activists. Then, new topics, not very traditional for foresters, entered the arenas of controversy, professional practices, and (more slowly) university curricula. In many cases, these were important and ancient issues that are just beginning to gain due recognition.

With the publication in August 2021 of the latest IPCC (International Panel on Climate Change) report, this issue reaches the highest level of political importance, including a recognition of the importance of natural forest ecosystems in regulating climates. Management practices and official standards have to adapt to these realities.

Many issues that have been gaining recognition in the forestry sector have to do with the social importance of forests, their productivity, their management, and the division of their benefits. This is well known in the field, although not always in universities. In February of this year, there was a Regional Consultation on Forest Education in Latin America, supported by FAO, IUFRO, and ITTO. Its interim report had concluded that "Forestry education in Latin America and the Caribbean is going through an important historical moment. On the one hand, there is a constant demand for applicants for technical, undergraduate, and postgraduate programs and on the other, international forest policies are increasingly demanding forestry professionals. However, the importance of forests in maintaining human well-being is not an issue that is covered within the programs and the forest is still conceived as a supplier of wood and its derived products.

This vision affects the image of the forestry professional and has caused other related programs to compete with forest programs. “This concern is nothing new. In 1968, just as I was managing tropical forests in Uganda, Mr. Jack Westoby, representing the FAO Forestry Department, in a speech on forest management goals, noted that: “Contrary to what many outsiders believe, forestry is not, in its essence, about trees. It is about people. It is about trees only so far as they serve the needs of people. “In other words, the “forestry” profession exists to ensure that forest resources, ecosystems, and values continue to sustain the needs and wills of the population in all its variety.

In this same year of 1968, I was managing tropical forests in Uganda. We use silvicultural treatments considered excellent by the foresters of several tropical governments in the 1950s and 1960s. We use arboricides to poison “undesirable” trees to favor “desirable” species: practices now considered totally incompatible with biodiversity conservation, and with today’s certification.

In the next half century, many more elements have been added to the duties of forest managers, elements not very considered in the old days. For example: gender equity. And the United Nations Guiding Principles on Business and Human Rights, adopted in 2011, finally provided a guide for the implementation of the 1948 Universal Declaration of Human Rights, a very relevant issue for timber companies in their dealings with workers, neighbors, communities, and indigenous peoples. And gender equity.

This approach is not about “social forestry.” It is rather a recognition that all forestry is social. Especially in Mexico, with its high proportion of forest resources on the lands of ejidos and communities. In some regions of the country, since the 1980s, these owners are achieving successful, productive, profitable, and even certified forestry activities and companies. Here in the north, in arid and semi-arid areas, there are multiple resources of commercial value such as candelilla, oregano, sotol, and ixtle, and multiple official norms, rules, and procedures: there is a lack of forestry initiatives so that owners can organize themselves to reduce their poverty and vulnerability.

These comments go as a form of celebration, of recognizing that the field of training and research has diversified in the faculty, and that this book highlights it. Also, to put on the table the new challenges of research, training, and labor market that the forestry sector demands with its growth.

Dr. Timothy Jasper Synnott Hillary

Senior leader of the project for the foundation of the Campus Linares of the Universidad Autónoma of Nuevo León

Contrary to what many outsiders believe, forestry is not, in its essence, about trees. It is about people. It is about trees only so far as they serve the needs of people.—Jack Westoby, FAO, 1968

Contents

Contributors

Irma Delia Enriquez Anchondo Instituto de Ciencias Biomédicas, Universidad Autónoma de Ciudad Juárez, Ciudad Juárez, Chihuahua, Mexico

Asunción Andreu-Soler Facultad de Ciencias Marinas, Universidad Autónoma de Baja California, Ensenada, Baja California, México

Faustino Camarena-Rosales Facultad de Ciencias, Universidad Autónoma de Baja California, Ensenada, Baja California, México

A. Carrillo-Parra Universidad Juárez del Estado de Durango, Instituto de Silvicultura e Industria de la Madera, Ciudad Universitaria, Durango, Durango, Mexico

Javier Isaac de la Fuente Universidad de Quintana Roo, División de Ciencias de la Salud, Chetumal, Quintana Roo, Mexico

Inés Yañez Díaz Laboratorio de Suelos, Facultad de Ciencias Forestales, Universidad Autónoma de Nuevo León, Monterrey, Nuevo León, Mexico

Fortunato Garza-Ocañas Laboratorio de Suelos, Facultad de Ciencias Forestales, Universidad Autónoma de Nuevo León, Monterrey, Nuevo León, Mexico
Universidad Autónoma de Nuevo León, Campus Linares, Facultad de Ciencias Forestales, Linares, Nuevo León, Mexico
Laboratorio de Micología, Facultad de Ciencias Forestales Universidad Autónoma de Nuevo León, Linares, Nuevo León, Mexico
Facultad de Ciencias Forestales, Universidad Autónoma de Nuevo León, Linares, Nuevo León, Mexico

Ricardo Valenzuela Garza Laboratorio de Micología, Departamento de Botánica, Escuela Nacional de Ciencias Biológicas, Instituto Politécnico Nacional, Mexico, Mexico

J. R. Goche-Telles Universidad Juárez del Estado de Durango, Facultad de Ciencias Forestales, Durango, Durango, Mexico

Gonzalo Guevara Guerrero Tecnológico Nacional de México, Instituto Tecnológico de Ciudad Victoria, Ciudad Victoria, Tamaulipas, Mexico

Erica Esmeralda Hernández-Hernández Facultad de Ciencias Forestales, Universidad Autónoma de Nuevo León, Linares, Nuevo León, Mexico

S. J. Hernández-Mata Universidad Autónoma de Nuevo León (UANL), Facultad de Ciencias Forestales, Linares, Nuevo León, Mexico

Wibke Himmelsbach Facultad de Ciencias Forestales, Universidad Autónoma de Nuevo León, Linares, Nuevo León, Mexico

León Esteban Ibarra-Garibay Universidad Nacional Autónoma de México, UMDI Juriquilla, Facultad de Ciencias, Laboratorio de Ecología de Artrópodos en Ambientes Extremos, Juriquilla, Querétaro, Mexico

Jesús García Jiménez Tecnológico Nacional de México, Instituto Tecnológico de Ciudad Victoria, Ciudad Victoria, Tamaulipas, Mexico

Edwin Garza López Facultad de Ciencias Forestales, Universidad Autónoma de Nuevo León, Linares, Nuevo León, Mexico

Karen Elisama Rivera Luna Laboratorio de Micología, Facultad de Ciencias Forestales Universidad Autónoma de Nuevo León, Linares, Nuevo León, Mexico

Humberto Quiroz Martínez Facultad de Ciencias Biológicas, Universidad Autónoma de Nuevo León, Laboratorio de Entomología, San Nicolás de los Garza, Nuevo León, Mexico

Miroslava Quiñónez Martínez Instituto de Ciencias Biomédicas, Universidad Autónoma de Ciudad Juárez, Ciudad Juárez, Chihuahua, Mexico
Departamento Forestal, Universidad Autónoma Agraria Antonio Narro, Saltillo, Coahuila, Mexico

Yesenia Marisol Mejorado Martínez Facultad de Ciencias Forestales, Universidad Autónoma de Nuevo León, Linares, Nuevo León, Mexico

Horacio Villalón Mendoza Faculty of Forestry Sciences, Autonomous University of Nuevo León, Linares, Nuevo León, Mexico
Facultad de Ciencias Forestales, Universidad Autónoma de Nuevo León, Linares, Nuevo León, Mexico

Moisés Ramírez Meráz Facultad de Ciencias Forestales, Universidad Autónoma de Nuevo León, Linares, Nuevo León, Mexico

Iván Alejandro Meza-Matty Instituto de Investigaciones Oceanológicas, Universidad Autónoma de Baja California, Ensenada, Baja California, México

Nelson Manzanares Miranda Centro de Investigación en Producción Agropecuaria, Universidad Autónoma de Nuevo León, Linares, Nuevo León, Mexico

Ivette Molinar Monsivais Instituto de Ciencias Biomédicas, Universidad Autónoma de Ciudad Juárez, Ciudad Juárez, Chihuahua, Mexico

Lourdes Garza Ocañas Departamento de Farmacología y Toxicología, Facultad de Medicina, Universidad Autónoma de Nuevo León, Monterrey, Nuevo León, Mexico

Artemio Carrillo Parra University Juárez del Estado de Durango, Instituto de Silvicultura e Industria de la Madera, Durango, Durango, Mexico
Universidad Juárez del Estado de Durango, Instituto de Silvicultura e Industria de la Madera, Ciudad Universitaria, Durango, Durango, Mexico

Rafael Peña-Ramírez Tecnológico Nacional de México, Instituto Tecnológico Superior de Irapuato, Irapuato, Guanajuato, Mexico

Regina Pérez-Domínguez Facultad de Ciencias Forestales, Universidad Autónoma de Nuevo León, Linares, Nuevo León, Mexico

Juan Pablo Pinzón Departamento de Botánica, Facultad de Medicina Veterinaria y Zootecnia, Universidad Autónoma de Yucatán, Mérida, Yucatán, Mexico

Angélica Judith Ocampo Ramos Faculty of Forestry Sciences, Autonomous University of Nuevo León, Linares, Nuevo León, Mexico

Juan Manuel Soto Ramos Facultad de Ciencias Forestales, Universidad Autónoma de Nuevo León, Linares, Nuevo León, Mexico

Elisama Rivera-Luna Facultad de Ciencias Forestales, Universidad Autónoma de Nuevo León, Linares, Nuevo León, Mexico

Gerardo Cuellar Rodríguez Laboratorio de Entomología, Facultad de Ciencias Forestales Universidad Autónoma de Nuevo León, Linares, Nuevo León, Mexico
Facultad de Ciencias Forestales, Universidad Autónoma de Nuevo León, Linares, Nuevo León, Mexico

Gorgonio Ruiz-Campos Facultad de Ciencias, Universidad Autónoma de Baja California, Ensenada, Baja California, México

David Reyes Ruvalcaba Instituto de Ciencias Biomédicas, Universidad Autónoma de Ciudad Juárez, Ciudad Juárez, Chihuahua, Mexico

José Isidro Uvalle Sauceda Facultad de Ciencias Forestales, Universidad Autónoma de Nuevo León, Linares, Nuevo León, Mexico

Mariana Solís-Mendoza Centro de Investigación Científica y de Educación Superior de Ensenada, Ensenada, Baja California, México

Michael Oswaldo Uitzil-Collí Post Grado en Biociencias, Laboratorio de Micología, Departamento de Botánica, Escuela Nacional de Ciencias Biológicas, Instituto Politécnico Nacional, Ciudad de Mexico, Mexico

Horacio Villalón-Mendoza Facultad de Ciencias Forestales, Universidad Autónoma de Nuevo León, Linares, Nuevo León, Mexico

Burghard Wittig Freie Universität Berlin, Berlin, Germany

María Haydeé Solano Zavala Instituto de Ciencias Biomédicas, Universidad Autónoma de Ciudad Juárez, Ciudad Juárez, Chihuahua, Mexico

Abbreviations

BWP	backwater pool
CCP	channel confluence pool
CF	Fixed carbon
GLD	glide
gmL	light
gmMa	shear resistance
gmN	soil nitrogen
gmR	soil reaction
HHV	Higher heating
LGR	low gradient riffle
LSP	lateral scour pool
MCP	mid-channel pool
PLP	pluge pool
RUN	run
SRN	step run
STP	step pool
VM	Volatile material

Conservation, Regeneration and Development of Species-Rich Meadows in Flooded Areas in Northwestern Germany

Burghard Wittig

Abstract Species rich meadows of the FFH-LRT 6510 belong to the most threatened habitat types in Germany. The regeneration and development of these plant communities have become very relevant for nature conservation. Species rich lowland meadows of the *Arrhenatheretum elatioris* have developed on former arable-fields, pastures, mown pastures, and temporarily fallow areas in the Verdener Wesertal. Today they are used mostly as meadows, mowed one or twice per year. All recent investigated sites were either pursued by nature conservation administration or they are still privately owned. The privately-owned sites are integrated in agri-environmental schemes. The recent meadows were compared with historical releves from the 60th of the last century. In the sixties, there were only small-scale stands used as mown pastures. Tall oatgrass-meadows developed from mown pastures and pure pastures. Now species rich-stands are especially established on former arable-fields that have been established spontaneously. Flooding must have had a strong influence on the spread of many species, although the stands are not flooded every year. The input of diaspores and soil dynamic (disturbance and/or aggradation) have formed new species combinations with high conservation value.

Keywords Pastures · Distribution · Meadows · Flooding · Agriculture

1 Introduction

The aim of the European Union's Fauna-Flora-Habitat Directive is to protect and protect wild species, their habitats, and the European network of these habitats. One of the many habitat types listed is the "lowland hay meadow" or FFH-LRT 6510 (EUROPEAN COMMISSION, DG ENVIRONMENT 2013). In Germany these meadows belonging to the *alliance of Arrhenatherion* are rich in flowers. According to Dierschke (1997), the most suitable sites of this type are found in deep, alkaline,

B. Wittig (✉)
Universität Bremen, Bremen, Germany
e-mail: bwittig@uni-bremen.de

F. Garza-Ocañas (ed.), *Sustainable Management of Natural Resources*, Earth and Environmental Sciences Library, https://doi.org/10.1007/978-3-031-33394-1_1

and nutrient-rich mineral soils in climatically favorable lowlands. Soils have a good water supply, but they are not very wet for a long time. Soils show a slight acidic to neutral reaction. They receive little or no fertilizer. The meadows were traditionally cut once or twice a year.

2 Objectives

It was very surprising to find new lowland hay meadows in the Verdener Wesertal 2006 (see in detail Wittig et. al 2019). The sites were investigated in 2006 and 2016. For less than twenty years, these species-rich communities have developed into ancient grass fields, mown pastures, and temporarily fallow areas, where they are mainly used as meadows today, mowed once or twice a year. In the Verdener Wesertal these meadows were rare also rare in the past in contrast to Central or Southern Germany. Research in the 60s of the last century of Hofmeister (1970) gives an idea, how these meadows looked in this region. But they occurred only on small scales on the southern slopes of dikes or on the hedge windshield. Today they cover sizes of several hectares, which is a success for nature conservation. These meadows were developed from mowed pastures, fallow land, and crop fields. What are the reasons for this development? What are the prospects for these meadows? The sites investigated (Fig. 1) are in the lowlands of Verden Weser or "Verdener Wesertal" (Landschaftsrahmenplan Verden 2008). Alluvial soils are characterized by weakly cool to slightly humid conditions. On average, the river meadows are flooded every year for up to 5 days (minimum 0 days, maximum 15 days, data evaluation period 1998 - 2017, NLWKN, Verden). All recently investigated sites were followed by the rural district of Verden or by the state or lower Saxony for conservation purposes and leased to farmers or are still privately owned. Farmers from the few privately owned sites participate in agri-environmental schemes.

3 Methodology

In the years 2004 and 2016, a total of 23 relevés were taken at different lowland prairie sites in the Verdener Wesertal (Fig. 1). The size of the area was 16 m^2. These relevés were compared with 28 historical relevés of Hofmeister (1970) with the same phytosociological type. A PCA was conducted. The mean values of the Ellenberg indicator for light, moisture, soil reaction and nitrogen (Ellenberg et al., 1992) and values for shear strength (Briemle & Ellenberg 1994) were calculated for a better understanding of the sites.

Statistical analyses were carried out with the help of R, version 3.4.2 (R Development Core Team 2017). Since the gradient lengths of the DCA analysis were less than 3, a PCA was performed for all vegetation relevés with transformed covers (Leyer & Wesche 2007). The Ellenberg mean values of the graphs were

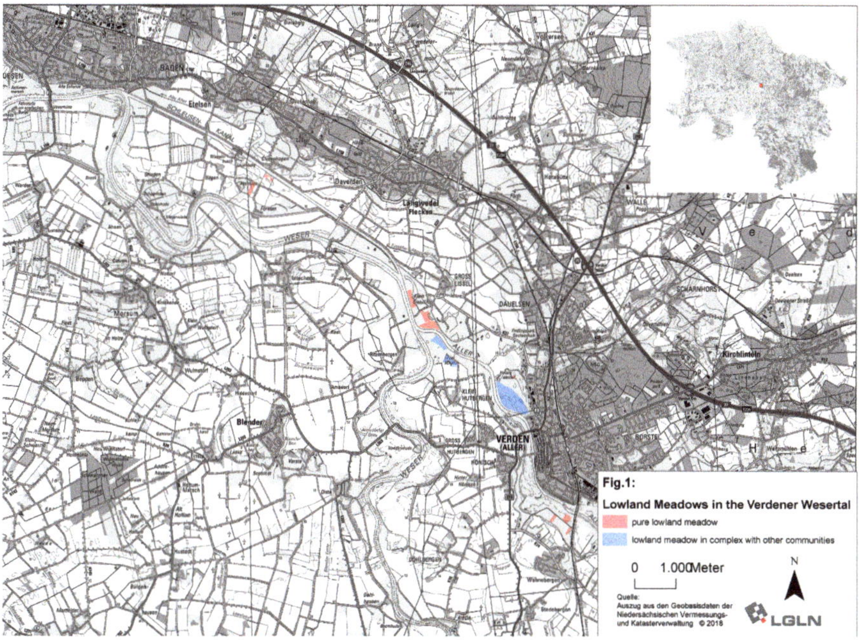

Fig. 1 Location of the oat-grass meadows investigated. The meadows of the Oat-grass are dominating in some fields. In other fields they occur in complexes with other types of vegetation. The complexes with sandy xeric grasslands and the community *Ranunculus repens - Alopecurus pratensis* - are located on the Maulohe directly in the north of Verden. Areas marked blue showing transitions to the community of *Ranunculus repens - Alopecurus pratensis*

adjusted post hoc to the results of the PCA sorting, and the significance of the effects of the variables was tested with Monte Carlo permutation (with 9999 permutations) separately for each variable.

Information on pre-use and current use comes from current land managers or from the archive's inspection at the Lower Nature Conservation Authority of the Verden district and the NLWKN, Lüneburg, the Lower Saxony Nature Administration. Information on the pre-use of the current sites was received from local farmers and the nature conservation administration (Landkreis Verden and NLWKN Lüneburg).

4 Results

Table 1 shows a highly shortened extract from a constancy table by Wittig et al. (2019). Only very characteristic species are mentioned. All relays belong to the oat grass meadow with the phytosociological name *Arrhenatheretum elatioris*. Both historical and recent relevés are divided into a Hypochaeris *radicata* Sub-Association and a Typical Sub-Association. The sub-association of

Table 1 Table of well-shortened constancy of historical and recent relevés of *the Arrhenatheretum* meadows in the Verdener Wesetal, H: Hofmeister 1970, AW: Verdener Wesertal 2006 and 2016 H1 and AW1: Subasociation of Hypochaeris radicata, H2 and AW2: Typical subsociation constancy: I = –20%, II = 20–40%, III = 40–60%, IV = 60–80%, V = 80–100%

Type	H1	AW 1	H2	AW 2
Number of relevés	5	5	23	15
Mean number of species	30.4	19.6	27.8	22.7
Arrhenateretum elatioris				
Arrhenatherum elatius	IV	V	IV	V
Bromus hordeaceus	IV	II	IV	II
Galium mollugo	IV	–	V	III
Crepis biennis	III	–	IV	III
Dactylis glomerata	V	–	V	II
Heracleum sphondylium	IV	–	V	II
Achillea millefolium	III	I	IV	III
Tragopogon pratensis	III	I	II	II
Veronica chamaedrys	III	–	II	II
Leucanthemum vulgare	II	–	I	III
Pimpinella major	–	–	I	–
Pastinaca sativa	II	–	III	–
Trisetum flavescens	III	–	I	IV
Indicators for dryness				
Galium verum	I	III	–	II
Hypochoeris radicata	IV	IV	–	+
Luzula campestris	II	III	II	–
Ranunculus bulbosus	IV	II	–	–
Armeria maritima subsp.	I	–	–	–
Allium oleraceum	I	–	–	–
Rumex acetosella	–	IV	–	–
Vicia tetrasperma	–	IV	–	+
Trifolium arvense	–	II	–	–
Dianthus deltoides	–	II	–	–
Grazing indicators				
Ranunculus repens	–	I	III	II
Bellis perennis	V	–	IV	II
Deschampsia cespitosa	–	–	III	–
Scorzonerioides autumnalis	II	–	II	–
Widespread meadow-plants				
Poa pratensis	V	I	V	IV
Rumex acetosa	IV	IV	V	V
Holcus lanatus	IV	V	IV	V
Plantago lanceolata	IV	IV	V	IV
Ranunculus acris	IV	I	IV	IV
Festuca rubra	IV	IV	III	III
Trifolium dubium	III	I	II	II

(continued)

Table 1 (continued)

Type	H1	AW 1	H2	AW 2
Agrostis capillaris	II	IV	I	IV
Anthriscus sylvestris	IV	–	V	I
Taraxacum officinale agg.	III	–	V	IV
Anthoxanthum odoratum	V	III	III	IV
Alopecurus pratensis	–	–	V	V
Vicia cracca	IV	–	IV	III
Lathyrus pratensis	II	–	II	II
Poa trivialis	IV	–	IV	IV
Festuca pratensis	III	–	IV	IV
Trifolium pratense	III	I	IV	IV
Stellaria graminea	IV	III	II	II
Trifolium repens	IV	I	III	III
Cerastium holosteoides	IV	–	V	III
Lolium perenne	II	–	II	II
Centaurea jacea	III	I	–	II
Cardamine pratensis	–	–	V	IV
Ranunculus auricomus	–	–	I	II

Hypochaeris radicata is characterized by an increased consistency of *Hypochaeris radicata, Luzula campestris, Ranunculus bulbosus* and other low-nutrient species. In the Typical Subasociation these species are absent. Compared to historical relevés, the much higher constancy of *Trisetum flavescens, Leucanthemum vulgare* and *Centaurea jacea* is remarkable. *Rumex thyrsiflorus* and *Rhinanthus minor* were not on the list of former relevés. However, *Galium mollugo, Crepis biennis* and *Bromus hordeaceus* were represented more consistently in the historical vegetation dataset. This is also the case for *Dactylis glomerata, Heracleum sphondylium, Anthriscus sylvestris, Pimpinella major, Pastinaca sativa, Daucus carota* were more frequent. Especially the indicator species for grazing such as *Deschampsia cespitosa, Trifolium repens* and *Bellis perennis* were more frequent in the historical investigation.

In the recent relevés *Deschampsia cespitosa* have no importance. The relevés in Hofmeister (1970) with high abundances of *Deschampsia cepitosa* are marked in Fig. 2 ("td"). The *Hypochaeris radicata Subassociation* is growing in more sandy and dry soils than the Typical Subassociaiton.

The weighted mean values of the nitrogen, reaction, and light indicator, as well as the weighted average mowing number indicate a significant influence of nutrient conditions and intensity of use on the composition of historical and current stand species. The first axis of the PCA (Fig. 2) correlates highly positively with the weighted indicator values of reaction number (rs = 0.96), nitrogen number (rs = 0.94) and mowing number (rs = 0.97), as well as negatively highly significant light weighted number (rs = −0.89).

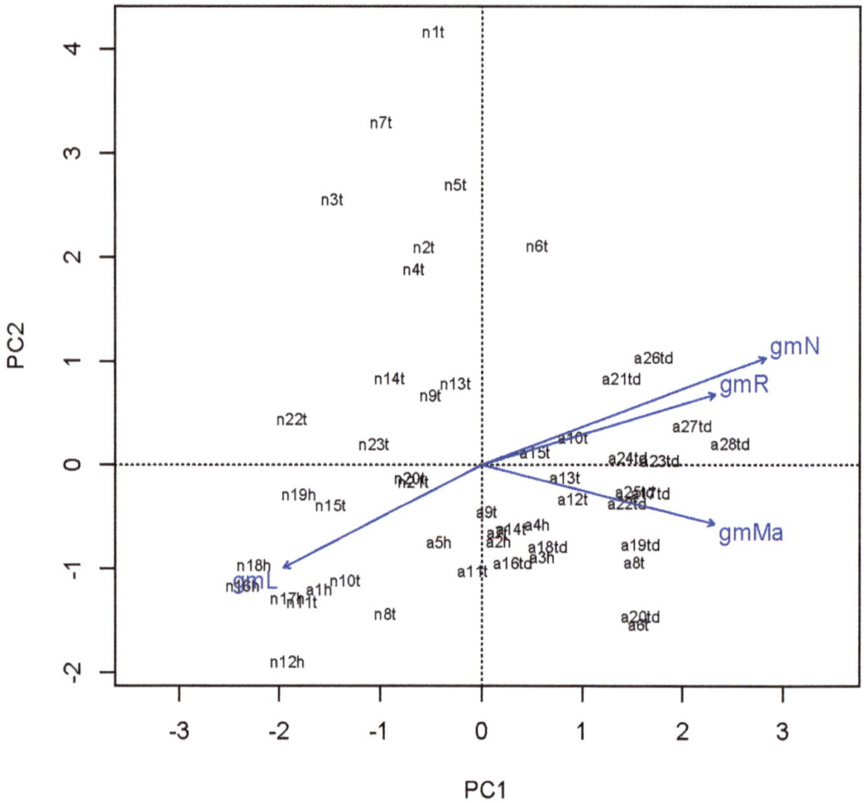

Fig. 2 PCA of historical and current relevés. PCA: 1. axis: eigenvalue: 25,474, explained variance 17.4%, 2. axis: eigenvalue: 14,879, explained variance 10.19%. Historical relevés a1 to a28, current relevés n1 to n23; Sub-association of Hypochaeris radicata, t: Typical sub-association, weighted average values of the indicator: gmN: soil nitrogen, gmR: soil reaction, gmL: light, gmMa: shear resistance

In general, the historical sites of Hofmeister (1970) tend to indicate more basic and nutrient-rich conditions. The average mowing rate is higher for the above sites. The currently most species-rich prairie (4.4 ha) was purchased by the rural district of Verden in 1989 for nature conservation. Previously, it was used as arable field and then had 2 years without any use (i.e. self-identification). Two other more species-rich fields were still fields in 1989 and were planted by herbs (Lolium perennial, Poa pratense). Most of the other land was previously used as standing or mowing grass.

The first cutting (mowing and meadows) took place in the Verden area usually in early June. Currently, the areas examined are generally mown by nature conservation requirements. The first cut is allowed after June 20 fertilization is prohibited. But a supply of nutrients takes place due to flooding. A second harvest is usually rare.

5 Discussion

A species rich in lowland grasslands of the alliance is *Arrhenatherion elatioris*, which covered entire fields, apparently not yet occurring in the area investigated in the 60s of the twentieth century. The fact that at the time of Hofmeister (1970) oat-grass meadows only appeared on small scales in the study area certainly has to do with the fact that grassland was mainly used as mown grass (first cut and then grass). The higher weighted average number of nitrogen and mowing from historical plant communities indicates that they have been used more intensively than current ones. Of these factors, grass has been the main source of influence in the past. An indication of this is the greater constancy of the grazing indicators *Bellis perennis* and *Trifolium repens*, but also of the *Deschampsia cespitosa* itself, which is quite poor tolerant cut (Briemle et al., 2002), in the old supports of Hofmeister (1970).

According to Ellenberg and Leuschner (2010), more frequent cuts and stronger fertilization, as well as temporary grazing, make oat meadows more fertile, but floristically poorer and ultimately "characterless." For the areas currently investigated, it can be assumed that the pool of species has survived in the ancient pastures mowed as adult plants and seeds. Other sites, where species survived, were riparian strips, roadsides and old dikes and remains. For example, the occurrences of *Trisetum flavescens*, *Galium album*, *Arrhenatherum elatius*, etc. are known from old sand dikes in Verden, which were grazed with sheep, for decades. Therefore, these species could spread after changing the use of grasslands from mown grass or pasture to a pure meadow.

Especially notable is the development of species-rich grasslands on ancient arable land (Fig. 3). The use of arable land lasted at least 10 years, before the use of grasslands began. Why are the first arable sites currently the most species-rich grasslands and could develop within about 25 years? The currently most species-rich area developed after the use of arable crops and self-identification. For oneself grassland diasporas can derive from the seed bank or the surrounding environment (Bosshard, 1999). Since the germination capacity of most grassland species is already depleted after 2–5 years (Bosshard, 1999), diasporas must come from the surrounding area. It can be assumed that flood events played an important role in the spread of many species and will continue to do so in the future, as all populations investigated are in the recent floodplain (Fig. 4). In addition, all ancient arable land is located downstream, north of all the investigated lands. The species must immigrate from the outside.

In addition to diaspora transport, disturbances caused by floods are also decisive for the success of species colonization (Bonn & Poschlod, 1998). Kleinschmidt & Rosenthal (1995) and Rosenthal (2006) show for wet meadows that flooding plays a role in seed dispersal of grassland species, they found seedlings of 81 species from 83 mud samples. Bonn & Poschlod, (1998) cite an unpublished work by Trottmann & Poschlod on small running waters rich in lime. Here too, some grassland species were found in the drift material. Vogt et al. (2006) were able to find more than 70 species in mud samples after the strong flooding of the Elbe in the summer of

Fig. 3 Oat grass meadow-species-rich near Verden-Eissel. The site was used as a arable field in 1989. After two years of abandonment, the site is used as a single-cut meadow (mowing after 20. June). Photo: José Müller

2002. They conclude that extreme flood events with their high energy are of great importance for the spread of many plant species in floodplains. Because most grassland species do not form a persistent seed bank and diaspore propagation is severely limited (Bakker et al., 1996), the spontaneous development of oat grasslands presented here is a great peculiarity and a special case today. Diaspore transport, soil dynamics (disturbance and/or landing) and extensive mowing have formed new combinations of species with high nature conservation value. It should be emphasized that the development or regeneration of species-rich oat grasslands without the active introduction of species is generally not possible in the most potentially suitable locations today (Bosshard, 2000; Kirmer & Tischew, 2014). The extent of grassland white species is mostly strong limited in intensive-used cultivated landscapes. For example, the transfer of hay from species-rich donor sites (nearby) to a prairie recipient is today an increasingly used method to regenerate species-rich meadows in that case, other methods are the introduction of plants by certified seed mixtures or the transplantation of plant soil with vegetation and seeds (Kirmer & Tischew, 2014). Locally decisive for the development of oat-grass meadows are typical floods, combined with moderate mowing (1–2 cuts). It will be interesting to see if these areas will become even richer in species in the coming years. For handling practices, it is essential to allow flexible and non-static mowing dates. From time to time, a previous harvest should take place, for example, at the end of May/beginning of June.

Fig. 4 Summer flood 2013 in the Verdener Wesertal. The arrows mark some of the sites investigated. Photo: Erich Schwinge

For the conservation of species-rich grasslands, the integration of nature conservation into agricultural farms is indispensable (Schumacher 2013, Schumacher et al., 2013). The area investigated is a center for the breeding and maintenance of horses. Hay from almost all the sites examined is used as feed for horses, including high-class and highly demanded breeding horses. Moderate use of hay and not the use of mulch or even the use of silage with early mowing are good prospects for developing lowland meadows. It is a large and rare exception today that species-rich oat-grass meadows can develop spontaneously from ancient intensively used fields in a relatively short time. With the existing species reservoir this is possible in recent floodplains with proper use. The spontaneous new development of oat grassland meadows is very important for the conservation of species and nature in view of the generally documented decline of these grasslands, especially in northwestern Germany.

References

Bakker JP, Poschlod P, Strykstra RJ, Bekker RM, Thompson K (1996) Seed banks and seed dispersal: important topics in restoration ecology. Acta Bot Neerl 45:461–490

Bonn S, Poschlod P (1998) Ausbreitungsbiologie der Pflanzen Mitteleuropas. - Quelle & Meyer: 404pp

Bosshard A (1999) Renaturierung artenreicher Wiesen auf nährstoffreichen Böden. Ein Beitrag zur Optimierung der ökologischen Aufwertung der Kulturlandschaft und zum Verständnis mesischer Wiesen-Ökosysteme. Dissertationes Botanicae 303: 194pp

Bosshard A (2000) Blumenreiche Heuwiesen aus Acker-und Intensivgrünland. Eine Anleitung zur Renaturierung in der landwirtschaftlichen Praxis. Nat Landsch 32:161–208

Briemle G, Ellenberg H (1994) Zur Mahdverträglichkeit von Grünlandpflanzen. Möglichkeiten der praktischen Anwendung von Zeigerwerten. Nat Landsch 69:139–147

Briemle G, Nitsche S, Nitsche L (2002) Nutzungswertzahlen für Gefäßpflanzen des Grünlandes. Schriftenreihe Vegetationskunde 38:203–225

Dierschke H (1997) Molinio-Arrhenatheretea (E1). Kulturgrasland und verwandte Vegetationstypen Teil 1: Arrhenatheretalia. Wiesen und Weiden frischer Standorte. – Synopsis der Pflanzengesellschaften Deutschlands 3: 1–74

Ellenberg H, Weber HE, Düll R, Wirth V, Werner W (1992) Zeigerwerte von Pflanzen in Mitteleuropa. 2. Aufl. - Scripta Geobotanica, vol 18. Goltze, pp 1–258

Ellenberg H, Leuschner C (2010) Vegetation Mitteleuropas mit den Alpen in ökologischer, dynamischer und historischer Sicht Ulmer, Stuttgart: 1344pp

European Commission. DG Environment (2013) Interpretation Manual of European Unit Habitats. 144pp. URL: http://ec.europa.eu/environment/nature/legislation/habitatsdirective/docs/Int_Manual_EU28.pdf. Accessed 04 May 2019

Hofmeister H (1970) Pflanzengesellschaften der Weserniederung oberhalb Bremens. - Dissertationes Botanicae 10: 116pp

Kirmer A, Tischew S (2014) Conversion of arable land to lowland hay meadows: what influences restoration success? In: Kiehl K, Kirmer A, Shaw N, Tischew S (eds) Guidelines for native seed production and grassland restoration. Cambridge Scholars Publishing, Cambridge, pp 118–140

Kleinschmidt C, Rosenthal G (1995) Samenbankpotential und Diasporenverdriftung in überschwemmten Feuchtwiesen. Kieler Notizen 23:40–44

Landkreis Verden (2008) Landschaftsrahmenplan. - URL: http://www.entera-online.com/013_verden/. Accessed 10 Nov 2018

Leyer I, Wesche K (2007) Multivariate Statistik in der Ökologie. Springer, Berlin, Heidelberg, p 221

R Development Core Team (2017) R: a language and environment for statistical computing. R Foundation for Statistical Computing, Vienna. Austria

Rosenthal G (2006) Restoration of wet grasslands – effects of seed dispersal, persistence and abundance on plant species recruitment. Basic Appl Ecol 7:409–421

Schumacher W (2013) Biodiversität extensiv genutzter Grasländer und ihre Erhaltung durch Integration in landwirtschaftliche Betriebe – Erfahrungen und Ergebnisse 1985–2012. In: Schröder S & Wider J (Eds): Agrobiodiversität im Grünland schützen. Bundesanstalt für Landwirtschaft und Ernährung: 70–99

Schumacher W, Trein L, Esser D (2013) Biodiversität von Magerrasen, Wiesen und Weiden am Beispiel der Eifel – Erhaltung und Förderung durch integrative Landnutzungen. Berichte der Reinhold Tüxen-Gesellschaft 25:56–71

Vogt K, Rasran L, Jensen K (2006) Evidence for hydrochorous short- and long- distance seed dispersal during an extreme flooding event. Basic Appl Ecol 7:422–432

Wittig B, Müller J, Mahnke-Ritoff A (2019) Talauen-Glatthaferwiesen im Verdener Wesertal (Niedersachsen). Tuexenia 39:249

Population Density of the Endemic Trout (*Oncorhynchus mykiss nelsoni*) and its Relationship with the Habitat in the Sierra San Pedro Mártir, Baja California, Mexico

Gorgonio Ruiz-Campos, Mariana Solís-Mendoza, Faustino Camarena-Rosales, Asunción Andreu-Soler, and Iván Alejandro Meza-Matty

Abstract The population density of the endemic trout (*Oncorhynchus mykiss nelsoni*) of the Sierra San Pedro Mártir, Baja California, Mexico was evaluated. The relationship of population density with physicochemical and structural variables of the habitat in three streams with differential altitude (553 to 2069 meters above sea level), between February 2014 and April 2017 was studied. A total of 105 habitat units were sampled covering a segment of 3461 meters of stream, representing ten types of habitat units, two in Arroyo La Grulla, six in Arroyo San Antonio de Murillos and eight in Arroyo San Rafael. The highest density (individuals/m^2) was recorded in Arroyo San Rafael (0.106 in February 2014 and 0.102 in March 2015). Trout density was dependent on the type of habitat unit, where individuals showed a strong preference for lateral erosion pools, step pools and current, and low-gradient rapids. The diversity of habitat units in the sampled streams was higher in Arroyo San Rafael in February 2014 (0.789 bits) and lower in Arroyo La Grulla in September 2016 (0.213 bits). Habitat variables of physicochemical type provided a greater explanation (63%) than those of a structural nature (39%) on the density of trout in streams, being the most important variables in decreasing order: dissolved oxygen, salinity, temperature, macrophytes coverage, substrate, altitude, and habitat

G. Ruiz-Campos (✉) · F. Camarena-Rosales
Facultad de Ciencias, Universidad Autónoma de Baja California, Ensenada, Baja California, México
e-mail: gruiz@uabc.edu.mx

M. Solís-Mendoza
Centro de Investigación Científica y de Educación Superior de Ensenada, Ensenada, Baja California, México

A. Andreu-Soler
Facultad de Ciencias Marinas, Universidad Autónoma de Baja California, Ensenada, Baja California, México

I. A. Meza-Matty
Instituto de Investigaciones Oceanológicas, Universidad Autónoma de Baja California, Ensenada, Baja California, México

F. Garza-Ocañas (ed.), *Sustainable Management of Natural Resources*, Earth and Environmental Sciences Library, https://doi.org/10.1007/978-3-031-33394-1_2

unit area. This investigation provides a baseline for future population monitoring of this endemic species and its habitats, which currently face threats from anthropogenic activities in the middle and upper Santo Domingo and San Rafael basins.

Keywords Trout of San Pedro Mártir · Population density · Habitat unit · Diversity of habitats · Environmental variables

1 Introduction

Habitat assessment in stream ecosystems represents the main source of information for the assessment of watershed conditions, the management of their aquatic resources (Dollof et al. 1993; Johnson et al. 2001) and the determination of the ecological integrity of these biotopes because of anthropogenic activities (Meehan 1991). In this type of studies, the habitat unit is structurally recognized as the portion of the stream that can be visually delimited based on its physiographic attributes (depth, substrate, slope, etc.) and hydrological attributes (current and flow velocity) (Bryant et al. 1992). In this context, at least 24 types of habitat units have been identified in the streams with salmonid fish (trout) of western North America, which are grouped into three major categories (pool, riffle and current) (USDA-USFS 1990).

Several abiotic and biotic factors have been considered of great importance in salmonid habitat quality, highlighting flow regime, water quality, habitat structure, energy sources (prey availability), biotic interactions, and site access (Meehan 1991; Spence et al. 1996; Cederholm et al. 2001). In the southernmost region of the native range of rainbow trout, *Oncorhynchus mykiss*, inhabits the subspecies *O. m. nelsoni* (Evermann, 1908), endemic to the western slope of the Sierra San Pedro Mártir, Baja California, Mexico (Ruiz-Campos and Pister 1995; Behnke 2002), whose genetic integrity has been favored by its peculiar geographical isolation (Camarena-Rosales et al. 2008; Abadía-Cardoso et al. 2016) and the still pristine conditions of their habitats (Ruiz-Campos 2017).

The life history of this endemic trout has been the subject of different studies regarding the composition and trophic ecology (Ruiz-Campos and Cota-Serrano 1992), reproductive biology (Ruiz-Campos 1993), age and somatic growth (Ruiz-Campos et al. 1997, 2016), population density (Ruiz-Campos 1993, 2017; Ruiz-Campos et al. 2014), natural repopulation (Ruiz-Campos 1989), home environment (Ruiz-Campos and Villalobos-Ramírez 1991), parasites (Valles-Ríos and Ruiz-Campos 1996), genetic variability (Camarena-Rosales et al. 2008), morphometric comparison with other native Mexican trout (Ruiz-Campos et al. 2003), and their phylogenetic relationships with other coastal rainbow trout of Southern California (Abadía-Cardoso et al. 2016). Although the current conservation status of *O. mykiss nelsoni* is considered stable (Ruiz-Campos and Pister 1995; Ruiz-Campos et al. 2003, 2014; Ruiz-Campos and Varela-Romero 2016), their populations and habitats could be threatened by anthropogenic disturbances at the headwaters of streams, mainly by the channeling and transfer of their flows, deforestation of riparian vegetation, grazing by livestock, mining extraction, the introduction of exotic species and global climate change that will eventually reduce habitat and altitudinal distribution at the metapopulation level (Ruiz-Campos 2017).

On this topic, Mitro et al. (2010) projected that a 2.4 °C increase in water in the Wisconsin region, USA, would result in a 94% and 33% reduction in habitat extent for brook trout (*Salvelinus fontinalis*) and brown trout (*Salmo trutta*) respectively. Meza-Matty et al. (2021) predicted by three models climatic (GFDL R30, HadCM3, and Mote) using increases of 0.75 and 2.0 °C in water for streams of distribution of the endemic trout (*O. mykiss nelsoni*) in the Sierra San Pedro Mártir, a reduction of 21–23% and 23–31% in the current altitudinal distribution range of this trout, during 2025 and 2050, respectively.

In the present study, we evaluated the population density of *O. mykiss nelsoni* in the different habitat units present in three streams with different elevations of the Sierra San Pedro Mártir and their relationship with environmental variables (physiographic, hydrometric, and physicochemical). The information generated here will be of critical es importance for the identification of the habitats with the greatest abundance of this trout and identify the associated environmental factors. All this as a basis for population monitoring of this subspecies and its habitats, as well as being a reference for future conservation, restoration, and habitat improvement programs.

1.1 Study Area

The Sierra San Pedro Mártir is the highest batholithic formation of the peninsular mountain range, which extends from southern California (USA) to the south of the Peninsula of Baja California, Mexico (O'Connor and Chase 1989; Barajas 2018), where the Picacho del Diablo is the highest (3096 masl). The western part of this mountain range presents an inclination towards the Pacific Ocean, while its eastern part is distinguished by a steep slope that, in an interval of 8 km, descends suddenly from 3096 to 530 m (Delgado-Argote 1991).

The surface hydrology of the Sierra San Pedro Mártir is characterized by a series of perennial currents at its headwaters, but they become intermittent in their middle and lower parts of their trajectories towards the Pacific Ocean during extreme drought conditions (Tamayo and West 1964). The main streams are, from north to south, San Rafael, San Telmo and Santo Domingo (Fig. 1). The mouths of all these streams are blocked to the sea by the formation of a sandy bar, except during the events of extraordinary flows after storms (Tamayo 1962).

The vegetation associated with the streams of this mountain range forms the riparian component, where it is possible to distinguish mesophilic tree forms such as *Populus fremontii, P. tremuloides, Platanus racemosa* and *Salix lasiolepis*; shrub forms such as *Baacharis salicifolia* in the banks of the streams along with herbaceous forms such as *Hydrocotyle* sp. and *Berula erecta*. Aquatic macrophytes are represented by emerging forms such as *Schoenoplectus californicus* and *Typha*

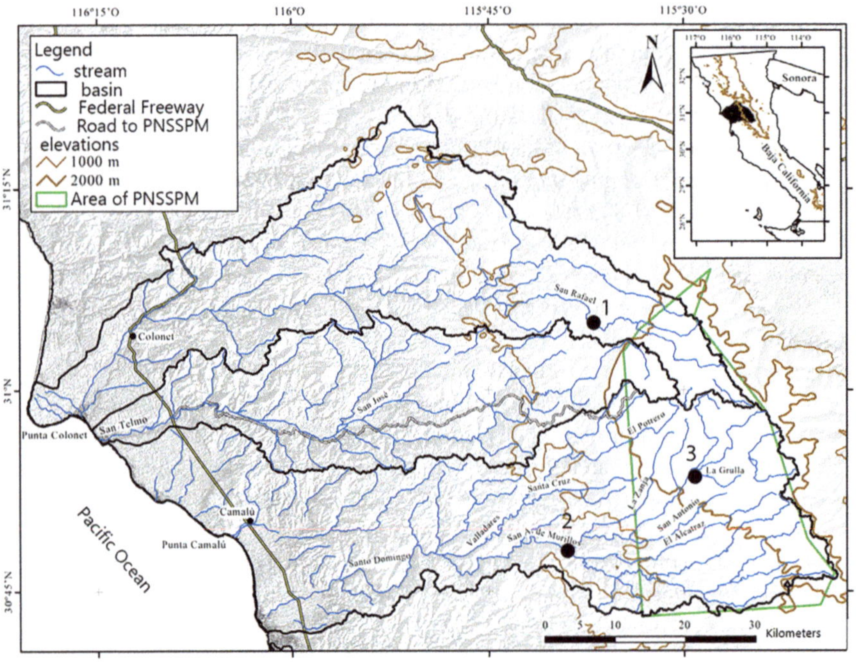

Fig. 1 Geographical location of the study area and sampling sites in the Sierra San Pedro Mártir, Baja California, Mexico. (1) Arroyo San Rafael at Rancho Mike's Sky, (2) Arroyo San Antonio de Murillos at Rancho San Antonio and (3) Arroyo La Grulla at La Grulla

domingensis, while macrophytes submerged by *Potamogeton natans, Nasturtium aquaticum* and *Ceratophyllum demersum* (Ruiz-Campos 2017).

2 Methodology

A total of nine sampling expeditions were conducted at three sites in the Sierra San Pedro Mártir, Baja California, Mexico, between February 2014 and April 2017 (Fig. 1). These sites are located at different elevations (Arroyo San Antonio de Murillos, 553 m; Arroyo San Rafael, 1230 m; and Arroyo La Grulla, 2069 m), which are within the known distribution range of the endemic trout *O. mykiss nelsoni* (Ruiz-Campos and Pister 1995; Ruiz-Campos et al. 2014).

At each site we selected a 169–780 m long stream segment for the identification of habitat units present, which were based on those described for salmonid fish in the Pacific drainage of North America (USDA-USFS 1990; Bain and Stevenson 1999). Each habitat unit was measured for its morphology (length and width, maximum and average depth, length and slope of the bank, dominant substrate), hydrometry (speed and discharge), water quality (temperature, dissolved oxygen, pH, salinity, conductivity, and total dissolved solids) and biological variables (macrophyte coverage). All these habitat variables were based on Dollof et al. (1993), Bain and Stevenson (1999), and Cornell et al. (2008).

The depth and current velocity in each habitat unit were measured in a cross-section of the stream at intervals of 30 cm, using a flexometer and a correntimeter (Model Swoffer 2100), respectively. The flow was calculated as Q= [W*D*V]*CF (Hynes 1972), where Q= discharge rate (m^3/s), W= average stream width, D= average depth, V= average current velocity and CF= friction constant (0.9). Physicochemical measurements were recorded using a Hydrolab Surveyor multianalyzer equipment (Hydrolab Co., Austin, Texas).

The density of trout was evaluated in each transect prior to the characterization of the habitat units. Each habitat unit was delimited at its ends with purse seines that were placed transversely into the streambed, thus preventing the entry or exit of individuals. Electrofishing equipment (LR-24 Smith-Root) was used for trout sampling, performing two sweeps on each habitat unit. All individuals captured in each habitat unit were kept alive in a 20-l container, then their total length was measured in millimeters, sex was determined, and finally they were released into their capture habitat unit.

The density of trout per unit of stream habitat was expressed as the number of individuals per square meter. A Chi-square (X^2) independence test was used to determine whether trout density is dependent on the type of habitat unit (Sokal and Rohlf 2012).

The diversity of habitat units (DHU) per sampled stream segment was calculated through an adaptation of Shannon's formula, such as DHU= -∑ ai/A*log ai/; where the number of species (s) was replaced by the number of habitats (a), and the abundance of each species (ni) by the area (m^2) occupied by each type of habitat unit (ai) in the sampled segment, and the total abundance of all species (N) by the total area (m^2) of the sampled segment (A). The values obtained were expressed in bits. To identify the main habitat variables that best explain the variation in trout density, a principal component analysis was used separately using the Statistica 7.0 package (StatSoft Inc Tulsa OK), one for physicochemical variables and one for structural variables. This analysis was performed only for expeditions conducted between February 2014 and August 2015, where the three streams were sampled.

3 Results

A total of 105 habitat units were sampled in three streams of the Sierra San Pedro Mártir, covering a segment of 3461 meters of streams in an altitude range of 553 to 2069 meters above sea level. Ten types of habitat units were identified, two in Arroyo La Grulla, six in Arroyo San Antonio de Murillos and eight in Arroyo San Rafael (Fig. 2).

The values recorded for the different habitat variables are given in Table 4 (Appendix).

Fig. 2 Classification of habitat units in the streams of the Sierra de San Pedro Mártir, Baja California, Mexico. Photographs by Gorgonio Ruiz-Campos

3.1 Trout Density in Arroyo La Grulla

On September 26, 2014, 50 individuals were collected in a segment of 262 m of stream with an average width of 4.43 m representing an area of 1060 m^2 and a population density of 0.047 individuals/m^2. The population densities in the two types of habitat units were 0.082/m2 in glide and 0.043/m^2 in mid-channel pools (Table 1).

In the sampling of August 9, 2015, 75 trout were collected through a segment of 169 m with an average width of 6.8 m, which represented an area of 1161 m^2 and a population density of 0.065 individuals/m^2. By type of habitat unit, the density of individuals/m^2 was 0.073 in mid-channel pool and 0.056 in glide (Table 1).

On September 17, 2016, 17 trout were caught through a 179 m segment of stream with an average width of 9.8 m, representing this segment an area of 1531 m^2 and a population density of 0.011 individuals/m^2. The density of individuals/m^2 in the habitat units of mid-channel pools and glides was 0.015 and 0.00, respectively (Table 1.).

On April 29, 2017, a total of 19 trout were caught in a 208 m segment of stream with an average width of 6.2 m, representing an area of 1521 m^2 and a population density of 0.013 individuals/m^2. The density of individuals/m^2 in the habitat units was 0.014 in glide and 0.011 in mid-channel pool (Table 1).

3.2 Trout Density in Arroyo San Antonio de Murillos

On May 3, 2014, 35 individuals were collected through a stream segment of 419 m in length and with an average width of 6.4 m, which represented a sampled area of 1664 m^2 and a population density of 0.021 individuals/m^2. At the habitat unit level, the densities (individuals/m^2) were step run (0.03) and run (0.03) > mid-channel pool (0.011) (Table 1).

During the sampling of April 13, 2015, 45 individuals were collected in a stream segment of 307 m with an average width of 4.5 m, where an area of 1353 m^2, providing a population density of 0.033 individuals/m^2. The density (individuals/m^2) by type of habitat unit was as follows: step run (0.070) > step pool (0.062) > low gradient riffle (0.045) > run (0.019) and mid-channel pool (0.000) (Table 1).

3.3 Trout Density in Arroyo San Rafael

In the sampling of February 23, 2014, 116 individuals were collected in a segment of 217 m in length and an average width of 4.0 m, representing a sampled area of 1097 m^2 and a population density of 0.106 individuals/m^2. The density ratio by habitat

Table 1 Trout density (individuals/m^2) by type of habitat unit in the streams of the Sierra de San Pedro Mártir, Baja California, Mexico, during the period February 2014 to April 2017

Sampling date	Stream site	Unit of habitat	Área (A) (m^2)	(% A)	Number of trout (N)	(% N)	Trout density per m^2
26 Sep. 2014	La Grulla	MCP	938.7	88.5	40	80	0.043
		GLD	121.4	11.5	10	20	0.082
		Total	1060.1		50		0.047
9 Aug. 2015	La Grulla	GLD	573.6	49.4	32	42.7	0.056
		MCP	587.8	50.6	43	57.3	0.073
		Total	1161.4		75		0.065
17 Sep. 2016	La Grulla	MCP	1118.8	73.1	17	100	0.015
		GLD	412.7	26.9	0	0	0.000
		Total	1531.4		17		0.011
29 Apr. 2017	La Grulla	GLD	710.6	46.7	10	52.6	0.014
		MCP	810.8	53.3	9	47.4	0.011
		Total	1521.4		19		0.013
3 May 2014	San Antonio	MCP	804.5	48.3	9	25.7	0.011
		RUN	233.2	14	7	20	0.030
		SRN	626.6	37.7	19	54.3	0.030
		Total	1664.2		35	100	0.021
13 Apr. 2015	San Antonio	RUN	411.9	30.4	8	17.8	0.019
		STP	176.6	13.1	11	24.4	0.062
		SRN	214.9	15.9	15	33.3	0.070
		CCP	370.4	27.4	6	13.3	0.016
		LGR	111	8.2	5	11.1	0.045
		MCP	68.4	5.1	0	0	0.000
		Total	1353.3		45		0.033
23 Feb. 2014	San Rafael	GLD	1043.9	95.2	108	93.1	0.103
		LGR	52.9	4.8	8	6.9	0.151
		Total	1096.9		116		0.106
21 March 2015	San Rafael	GLD	939.4	37.3	5	16.7	0.005
		SRN	196.6	7.8	1	3.3	0.005
		LGR	165.7	6.6	3	10	0.018
		RUN	929.6	36.9	8	26.7	0.009
		STP	286.3	11.4	13	43.3	0.045
		Total	2517.6		30		0.012
24 May 2015	San Rafael	PLP	16.2	1.3	2	8.7	0.124
		RUN	998.4	80	16	69.6	0.016
		BWP	18.4	1.5	0	0	0.000
		SRN	72.4	5.8	3	13	0.041
		LSP	9.1	0.7	1	4.3	0.110
		STP	95.2	7.6	1	4.3	0.011
		LGR	38.4	3.1	0	0	0.000
		Total	1248.1		23		0.018

MCP mid-channel pool, *GLD* glide, *RUN* run, *SRN* step run, *STP* step pool, *CCP* channel confluence pool, *LGR* low gradient riffle, *PLP* pluge pool, *BWP* backwater pool, and *LSP* lateral scour pool unit type (individuals/m^2) in order of importance was low gradient riffle (0.151) > glide (0.103) (Table 1).

In the sampling of March 21, 2015, 30 individuals were collected through a stream segment of 780 m in length and an average width of 2.9 m, covering an area of 2518 m^2 that contributed a population density of 0.012 individuals/m^2. By type of habitat unit, the density (individuals/m^2) in decreasing order was step pool (0.045) > low gradient riffle (0.018) > run (0.009) > step run (0.005) and glide (0.005) (Table 1).

On May 24, 2015, 23 trout were caught in a stream transect of 439 m in length and an average width of 2.7 m that covered an area of 1248 m^2 and recorded a population density of 0.018 individuals/m^2. By type of habitat unit, the density of trout (individuals/m^2) in decreasing order was: pluge pool (0.124) > lateral scour pool (0.110) > step run (0.041) > run (0.016) > step pool (0.011) > backwater pool and low gradient riffle (0.000) (Table 1).

3.4 *Comparison of Trout Density and Habitat Diversity Between Streams*

At the level of sampling sites, the highest densities of individuals/m^2 were recorded in the streams La Grulla (0.065 in August 2015 and 0.047 in September 2014) and San Rafael (0.106 in February 2014) (Fig. 3a). The lowest density of individuals/m^2 occurred in Arroyo La Grulla with a value of 0.011 in September 2016 (Fig. 3a). In Arroyo San Antonio the density of individuals/m^2 was lower (0.021) in May 2014 and higher (0.033) in April 2015 (Fig. 3a).

Trout density was dependent on the type of habitat unit ($X^2 = 298.6$, 8 d.f., $P < 0.01$), where individuals showed a strong preference for pool habitat units by natural damming, lateral erosion pool, and low gradient rapid.

Using the habitat diversity index proposed here, the diversity of habitat units in the streams sampled in the study period oscillated both spatially and temporally.The greatest diversity of habitat units was recorded in Arroyo San Rafael in February 2014 with a value of 0.789 bits, while the lowest diversity corresponded to Arroyo La Grulla (0.213 bits) in September 2016 (Fig. 3b). The density of individuals in the three streams was independent of the diversity of habitat units identified ($r = 0.09$, 7 d.f., $P > 0.05$).

3.5 *Relationship of Trout Density and Habitat Variables*

The analysis of main components applied to the physicochemical variables of the habitat and the trout population that were recorded during the sampling events from February 2014 to August 2015, using as supplementary variables the density and

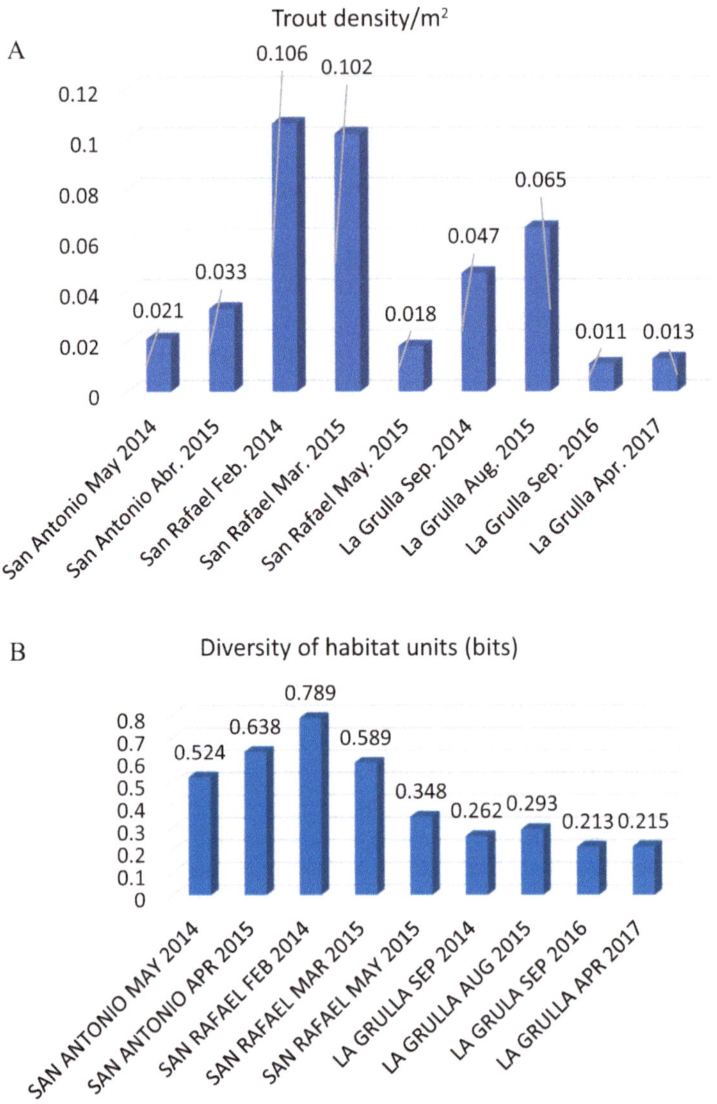

Fig. 3 Trout density/m^2 (**a**) in streams of the Sierra San Pedro Mártir, Baja California, Mexico during February 2014–April 2017, and diversity of habitat units (**b**) in these same streams

size of trout, the factors 1 (39.2%) and 2 (23.8%) explained of combined manner 63% of the observed variation (Fig. 4).

The physicochemical variables that contributed the most to the variation of trout density in factor 1 were those related to dissolved oxygen (0.290), salinity (0.274), temperature (0.225), and pH (0.183); while factor 2 were conductivity (0.456), pH (0.214), temperature (0.176), and dissolved oxygen (0.110) (Table 2).

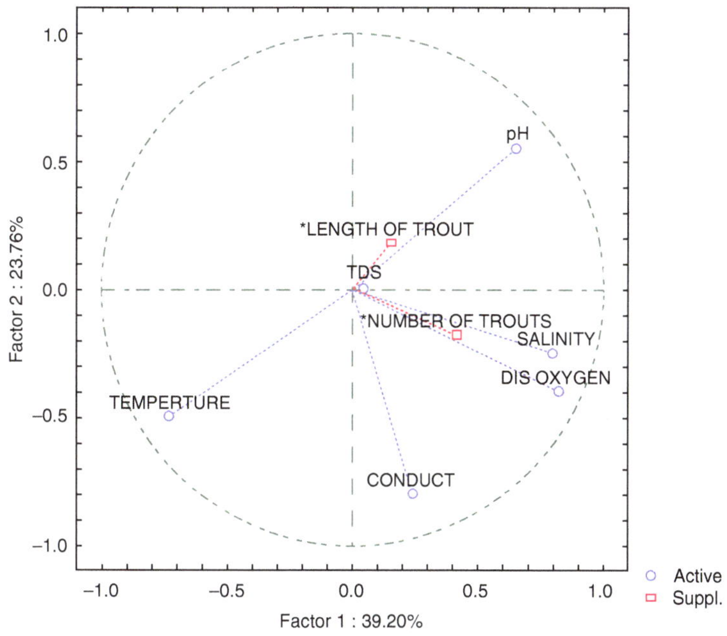

Fig. 4 Projection of active physicochemical variables and supplementary variables (number and length of trout) on the factor-plane 1 and 2, derived from principal component analysis for streams of distribution of *Oncorhynchus mykiss nelsoni*, in the Sierra San Pedro Mártir, Baja California, Mexico

Table 2 Contribution of physicochemical variables of the habitat of the trout *Oncorhynchus mykiss nelsoni* in the streams of the Sierra San Pedro Mártir, Baja California, Mexico, based on correlations using as supplementary variables the number and average size of trout

Physicochemical variables	Factor 1	Factor 2	Factor 3	Factor 4	Factor 5	Factor 6
TEMPERATURE	**0.225363**	**0.176469**	0.001073	0.035869	0.548699	0.012527
SALINITY	**0.273908**	0.043935	0.031567	0.276429	0.031819	0.342342
pH	**0.182951**	**0.213707**	0.013747	0.179755	0.409820	0.000020
DISSOLVED OXYGEN	**0.290530**	**0.110233**	0.000055	0.030139	0.000002	0.569041
CONDUCTIVITY	0.025973	**0.455593**	0.015090	0.428091	0.006069	0.069184
TDS	0.001275	0.000063	0.938467	0.049717	0.003591	0.006887

The values in bold were the most explanatory for the factors 1 and 2

On the other hand, the analysis of main factors for the structural variables of the habitat and the trout population recorded during the sampling but using as a supplementary variable the density of trout (number), explained through two factors combined 39.06% of the total variation observed (Fig. 5).

The structural variables of the habitat that contributed the most to explain the variation in trout density in factor 1 were macrophyte coverage (0.211), substrate

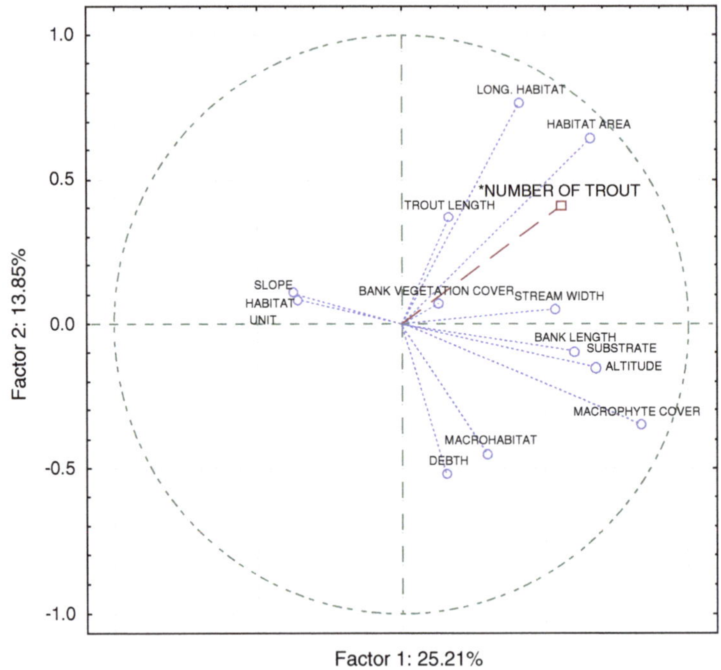

Fig. 5 Projection of active structural variables and the supplementary variable (number of trout) on the factor-plane 1 and 2, derived from principal component analysis for streams of distribution of *Oncorhynchus mykiss nelsoni*, in the Sierra San Pedro Mártir, Baja California, Mexico

(0.140), altitude (0.139) and habitat unit area (0.132); while factor 2 were length of habitat (0.324), habitat area (0.229), stream depth (0.149), and type of macrohabitat (0.112) (Table 3).

4 Discussion

4.1 *Population Density and Its Relationship to Habitat*

This study is the first of its kind for Mexico, evaluating the density of rainbow trout and its relationship to habitat in mountain streams. The average density estimated in the present study for the endemic rainbow trout (*Oncorhynchus mykiss nelsoni*) of the Sierra de San Pedro Mártir was 0.039 individuals/m^2 (range of 0.011 to 0.106 individuals/m^2), is within the reported range of 0 to 4.2 individuals/m^2 for 313 mountain streams in western North America (Platts and McHenry 1988). In this sense, the values obtained in our study on the observed density of the trout *O. mykiss nelsoni* are much lower than those recorded for brown trout (*Salmo trutta*) and rainbow in the eco regions of the Rocky Mountains (0.55/m^2), Gila Mountain (0.39/m^2), Pacific

Table 3 Contribution of structural variables of the habitat of the trout *Oncorhynchus mykiss nelsoni* in the streams of the Sierra San Pedro Mártir, Baja California, Mexico, based on correlations using as a supplementary variable the average number of trout

Structural variables	Factor 1	Factor 2	Factor 3	Factor 4	Factor 5
ALTITUDE	**0.138912**	0.012657	0.042208	0.125398	0.104269
HABITAT UNIT	0.038170	0.003675	0.100325	0.063024	0.020573
MACROHABITAT TYPE	0.027411	**0.111701**	0.156895	0.128488	0.026129
LENGTH OF HABITAT	0.051968	**0.323709**	0.021702	0.064078	0.016379
STREAM WIDTH	0.086940	0.001466	0.031803	0.294870	0.096279
HABITAT AREA	**0.132214**	**0.228744**	0.021521	0.008965	0.003087
STREAM DEPTH	0.007471	**0.149352**	0.171177	0.098641	0.032354
SUBSTRATE TYPE	**0.140083**	0.012772	0.080501	0.043976	0.052095
BANK LENGTH	0.110498	0.004983	0.008208	0.004213	0.152634
STREAM SLOPE	0.041589	0.005553	0.003470	0.056974	0.297643
BANK VEGETATION COVER	0.004994	0.002753	0.358122	0.085432	0.001056
MACROPHYTES COVER	**0.211736**	0.067615	0.002663	0.000485	0.022085
TROUT LENGTH	0.008014	0.075019	0.001406	0.025458	0.175419

The values in bold were the most explanatory for the factors 1 and 2

($0.29/m^2$), Columbia ($0.22/m^2$), Sierra Nevada ($0.16/m^2$), Intermontane ($0.40/m^2$), and southern California ($0.119–0.362/m^2$) (Platts and McHenry 1988; Barabe 2021); however, it was very similar to the density reported for the Colorado Plateau ($0.07/m^2$, Leiner 1995) and Saghen Creek, California ($0.01/m^2$, Decker and Erman, 1992). Similarly, the density of *O. mykiss nelsoni* is within the range of 0.008–0.348 trout/m^2 recorded by Leiner (1995) in 15 mountain streams of New Mexico, USA, as well as such as those of $0.02–0.17/m^2$ in 17 streams of the Sierra Nevada, California (Knapp and Dudley, 1990).

The density of trout in streams is dependent on different abiotic and biotic factors that influence spatio-temporally the productivity and carrying capacity of individuals in habitats (Hall and Knight 1981). Hynes (1972) determined that the abiotic factors that control survival in river habitats are water temperature, current velocity, exhaust coverage, and discharge regime. Other authors such as Lewis (1969) and Rinne (1982) identified that the volume of the pool is significantly correlated with the density of trout in populations of Montana and New Mexico (USA), respectively. For its part, discharge (flow) has been pointed out as a variable that explains the variation in density in brown trout in Wisconsin (White 1975).

Other studies have identified coverage as the limiting factor of trout population density (Hunt 1974; Binns and Eiserman 1979; Wesche 1980), or factors such as creek depth (Stewart 1970), the amount of organic debris or debris (Sedell et al. 1982) and invertebrate biomass (Murphy 1979).

The lower density of trout in the Sierra San Pedro Mártir could also be explained by the lower productivity in invertebrate biomass that has been recorded for these streams 14.46–27.68 g/m^2 and 0.59–23.87 g/m^2 (Solís-Mendoza 2016), and which are below those of 54.32–72.06 g/m^2 for other streams with rainbow trout in North America (cf. Surber 1937; Berg and Hellenthal 1992).

In the scenario of the streams of the SSPM and having as main actor the endemic trout, the differences in the density values of individuals crossing a spatial scale (sites of different elevation) and temporal (seasons of the year) can be explained statistically more by the physicochemical component (63%) than by the structural component of the habitat (39%). In this sense, the physicochemical and structural variables that best explain the differences in trout density between sites or streams were the level of dissolved oxygen, salinity, type of substrate, temperature, length, and area of habitat and macrophyte cover; all these factors are quite associated with the type of habitat unit, being more abundant trout in those habitat units such as lateral erosion pools, low gradient riffles and pluge pools.

As can be measured from the above, it is difficult to consider a single factor as the main limitation of the density of the trout population, rather, it is a combination of environmental variables that negatively or positively influence, on a spatio-temporal scale, on the trout population in a region (Platts and McHenry 1988).

4.2 *Historical and Current Comparison of Population Density*

In the La Grulla stream the density of trout/m^2 ranged from 0.008 (September 2016) to 0.064 (August 2015), the latter being very similar to that registered on this site (0.05/m^2) during October 2, 1994 (Ruiz-Campos 2017). On the other hand, the density of trout in the Arroyo San Antonio de Murillos in May 2014 (0.013/m^2) and April 2015 (0.033/m^2), were quite low compared to that previously recorded of 0.185/m^2 on October 1, 1995 (Ruiz-Campos 2017). In the case of Arroyo San Rafael, the observed trout density varied between 0.011/m^2 (September 2016) and 0.106/m^2 (February 2014), the latter higher than the average of 0.047/m^2 registered in this same locality during the period March 1987-August 1989 (Ruiz-Campos 1993; Ruiz-Campos and Pister 1995).

Although trout density varied between sampling localities and within each locality, it can be said that the density of individuals has remained relatively stable for at least the last 30 years (1987 to 2017).

4.3 *Physicochemical Factors*

The physicochemical characteristics of the water in the SSPM streams are dependent on the dynamics of meteorological factors and the chemical nature of the rocks in the drainage areas (Ruiz-Campos 1993). Temperature records at 30-minute intervals at the three sampling sites indicate the lowest values (1.11 °C) at Arroyo La Grulla (2035 masl) on February 1–2, 2016, and the highest (30.8 °C) at Arroyo San Antonio (560 masl) on August 15, 2014 (Ruiz-Campos 2017; Meza-Matty et al. 2021). As a primary source of comparison for pH, there are data in the Arroyo de San Rafael collected during the period from February 1989 to October 1992 (Ruiz-Campos 1993). In this study, an average pH variation of 8.0 to 8.7 was recorded, which is slightly less alkaline than that recorded in the present study from 8.4 to 9.98. The slight increase especially in the sampling of May 2015 (10.1) could be due to the effect of erosion of the banks of the stream due to the effect of trampling by cattle that use these sites to graze on the vegetation of the banks and the riparian zone, which causes a greater contribution of material from the soil to the stream channel that increases the amount of dissolved cations (Calcium and Magnesium) in the water and, therefore, the pH.

4.4 Frequency of Trout in Habitat Units

The highest densities of trout in the study streams in the Sierra San Pedro Mártir corresponded to those habitats of rare presence and of little covered surface, these being belonging to the macrohabitat of pool. Ruiz-Campos (1993) pointed out that trout prefer pools, which was reflected in a greater abundance in this type of habitat, especially in the Arroyo La Grulla (57.3%) in August 2015, where the habitat unit of the mid-channel pool is characterized by a high coverage of macrophytes. In the San Antonio de Murillos stream (Rancho San Antonio locality), in the May 2014 sampling, the highest density was obtained in the step run habitat unit (54.3%). At a global level in the Sierra San Pedro Mártir, the highest density of trout occurred in the habitat units of pools such as mid-channel, step, and natural damming (pluge), since these habitats are used as temporary thermal shelters during the day (Baltz et al. 1991), since being deeper they are thermally more stable and with a greater amount of dissolved oxygen, thus promoting a lower metabolic expenditure in individuals (Jonsson et al. 1991).

4.5 Recommendations on Management and Conservation

The current habitat of the trout of San Pedro Mártir is considered well preserved (Ruiz-Campos 2017), which is a consequence of the remoteness of the distribution localities of this endemic trout to human settlements and the difficulty of access to them. However, in recent years, the anthropogenic impact of agricultural activity has been increasing significantly in the lower parts of the basins in the coastal valleys of Colonet, San Telmo, Camalú and Vicente Guerrero, which demand water supplies for agricultural crops that are provided by wells built in the beds of streams, or by the channeling of water (piping) from the upper part of the basins. This situation could be complicated in the eventual use of water from the localities of current trout distribution as would be the case of the San Antonio de Murillos, El Potrero, La Zanja and San Rafael streams. Therefore, it is essential that in the short-term appropriate prevention measures are taken for the integral conservation of the habitats where this trout and other native aquatic forms are distributed.

The present study serves as a baseline for the future monitoring of the population density of the trout of San Pedro Mártir and the quality and quantity of their habitats, this being an indispensable source of information to elucidate the possible environmental factors that modulate the population density. That is because the monitoring will allow to know the patterns of abundance of the trout in the medium and long term, for which it is recommended that inventories of habitat units and population density are carried out, at least every three years (Hunter 1990).

Acknowledgments Our thanks to the different people who supported the field sampling in the Sierra San Pedro Mártir during the study period, especially Edgar Octavio Flores Hernández, José Guadalupe Sánchez Medina, Juan Corral Ávila, Javier Marco Barba, Itauhki Solís Mendoza, Bertha Paulina Díaz Murillo, Aldo Antonio Guevara Carrizales, Carlos Alonso Ballesteros Córdova, Orlando Acosta Seaman, Germán Ruiz-Cota, Marco Antonio Pimentel Lizárraga, David Ceseña Gallegos, Sarai Aranda Arreola, Gerardo Rivas Lechuga and Sergio Alejandro Celaya Delgado. To ranchers Rolando Arce Pechis and Indalecio Salazar for logistical support in transporting equipment and personnel with mules and horses to the study sites. This study received financial support from the 17th Internal Call for Research Projects via project "Characterization of aquatic habitats of the Sierra San Pedro Mártir, Baja California, and its relationship with the density and population structure of the endemic trout *Oncorhynchus mykiss nelsoni* (Evermann). Likewise, to the National Commission of Protected Natural Areas (CONANP), through the project PROCER/CCER DRPBCPN/ 03/2016 "Evaluation of the current state of conservation of six species of endemic vertebrates of the PN Sierra de San Pedro Mártir, Baja California". Finally, to the Academic Body "Studies Related to Biodiversity" of the Autonomous University of Baja California.

Appendix

Table 4 Mean and standard deviation (SD) of habitat variables measured in three streams of the Sierra San Pedro Mártir, Baja California, Mexico, during February 2014 to April 2017

Stream		Date	Depth (m)	Curr. Vel. (m/s)	Flow (m^3/s)	Slope (°)	Cover by Macrophytes	Temp. (°C)	Salinity (ppt)	pH	Diss. Oxyg. (mg/l)	Conduct. (mS/cm)	TDS (g/l)
SRS	Mean	Feb. 2014	0.13	0.38	0.16	20	1	11.5	0.16	9.4	11.2	0.30	0.13
	SD	Feb. 2014	0.04	0.08	0.02	7	0	0.7	0.03	0.6	0.3	0.09	0.00
SAR	Mean	May. 2014	0.26	0.17	0.13	15	1	24.8	<0.1	7.1	6.6	0.32	0.15
	SD	May. 2014	0.16	0.17	0.09	6	0	4.4	0.0	0.0	1.5	0.01	0.00
LSG	Mean	Sep. 2014	0.40	0.16	0.06	13	71	17.5	0.1	8.7	5.6	0.19	0.14
	SD	Sep. 2014	0.33	0.29	0.08	7	19	1.3	0.0	0.3	1.0	0.06	0.01
SRS	Mean	Mar. 2015	0.18	0.34	0.13	20	1	16.9	<0.1	8.4	6.1	0.27	0.16
	SD	Mar. 2015	0.10	0.14	0.07	13	0	3.3	0.0	0.3	0.2	0.04	0.14
SAR	Mean	Abr. 2015	0.20	0.25	0.18	25	10	17.3	<0.1	8.8	5.3	0.14	0.13
	SD	Abr. 2015	0.09	0.13	0.23	20	16	1.9	0.0	0.3	0.5	0.05	0.06
SRS	Mean	May. 2015	0.24	0.45	0.21	33	1	15.8	0.1	9.9	6.8	0.28	0.18
	SD	May. 2015	0.09	0.15	0.07	15	0	0.8	<0.1	0.1	0.7	0.00	0.00
LSG	Mean	Aug. 2015	0.20	0.01	0.40	12	93	19.7	0.10	7.4	4.4	0.22	0.09
	SD	Aug. 2015	0.14	0.00	0.27	5	3	0.9	0.00	0.2	0.3	0.20	0.04
LSG	Mean	Sep. 2016	0.18	0.04	0.05	15	57	15.4	0.2	8.3	6.3	0.33	0.21
	SD	Sep. 2016	0.06	0.04	0.03	3	26	1.6	0	0.2	0.8	0.06	0.04
LSG	Mean	Apr. 2017	0.14	0.14	0.07	15	96	16.3	0.21	6.8	7.8	0.42	0.27
	SD	Apr. 2017	0.04	0.11	0.06	2	2	0.4	0.0	0.1	0.8	0.06	0.03

SRS San Rafael stream, *LGS* La Grulla stream, *SAR* San Antonio stream

References

Abadía-Cardoso A, Pearse D, Jacobson S, Marshall J, Dalrymple D, Kawasaki F, Ruiz-Campos G, Garza JC (2016) Population genetic structure and ancestry of steelhead/ rainbow trout (Oncorhynchus mykiss) at the extreme southern edge of their range in North America. Conserv Genet 17:675–689

Bain MB, Stevenson NJ (1999) Aquatic habitat assessment: common methods. Fisheries Society, Bethesda, MD, p 216

Baltz DM, Vondracek B, Brown LR, Moyle PB (1991) Seasonal changes in microhabitat selection by rainbow trout in a small stream. Trans Am Fish Soc 120:166–176

Barabe RM (2021) Population estimates of wild rainbow trout in a remote stream of southern California. California Fish Wildl J 107:21–32

Barajas AM (2018) El origen. Págs. 23-36, En Semeel Jak: historia natural y cultural de la Sierra de San Pedro Mártir (E. Garduño y E. Nieblas, coords.). Tirant Lo Blanch, Cd. de México. 306pp

Behnke RJ (2002) Trout and salmon of North America. The Free Press, New York, p 384

Berg MB, Hellenthal RA (1992) Secondary production of Chironomidae (Diptera) in a north temperate stream. Freshw Biol 25:497–505

Binns NA, Eiserman FM (1979) Quantification of fluvial trout habitat in Wyoming. Trans Am Fish Soc 108:215–228

Bryant MD, Wright BE, Davies BJ (1992) Application of a hierarchical habitat unit classification system: stream habitat and salmonid distribution in Ward Creek, Southeast Alaska. United States Department of Agriculture, Forest Service-Pacific Northwest Research Station Research Note PNW-RN-508. March 1992. 18pp

Camarena-Rosales F, Ruiz-Campos G, De La Rosa-Vélez J, Mayden RL, Hendrickson DA, Varela-Romero A, García-De León FJ (2008) Mitochondrial haplotype variation in wild trout populations (Teleostei: Salmonidae) from northwestern Mexico. Rev Fish Biol Fish 18:33–45

Camarena-Rosales F, Ruiz-Campos G, De La Rosa-Vélez J, Mayden RL, Hendrickson DA, Varela-Romero A, García-De León FJ (2008) Mitochondrial haplotypes variation in wild trout populations (Telesotei: Salmonidae) from Northwestern Mexico. Rev Fish Biol Fish 18:33–45

Cederholm CJ, Johnson DH, Bilby R, Dominguez L, Garrett A, Graeber W, Greda EL, Kunze M, Palmisano J, Plotnikoff R, Pearcy B, Simenstad C, Trotter P (2001). Pacific salmon and wildlife-ecological contexts, relationships, and implications for management. 628–684, En D.H. Johnson and T.A. O'Neil (eds.). Wildlife-habitat relationships in Oregon and Washington. Oregon State University Press, Corvallis, OR

Cornell JE, Gutiérrez M, Wait AD, Rubio-Arias HO (2008) Ecological characterization of a riparian corridor along the Río Conchos, Chihuahua, Mexico. Southwest Nat 53:96–100

Delgado-Argote LA (1991) El Plutón de San Pedro Mártir: una visita geológica. In: Lazcano C (ed) Memoria III Semana de la Exploración y la Historia: Sierra de San Pedro Mártir. Universidad Autónoma de Baja California, Ensenada, Baja California, pp 25–27

Dollof CA, Hankin DG, Reeves GH (1993) Basinwide estimation of habitat and fish populations in streams. United States Department of Agriculture-Forest Service, Southeastern Forest Experiment Station, General Technical Report SE-83. 25pp

Hall JD, Knight JJ (1981) Natural variation in abundance of salmonid populations in streams and its implications for design of impact studies. EPA-600/53-81-021. Corvallis, OR: U.-S. Environmental Protection Agency. 5pp

Hunt RL (1974) Annual production by brook trout in Lawrence Creek during eleven successive years. Tech. Bull. 82. Wisconsin Department of Natural Resources, Madison, Wisconsin. 29 pp

Hunter CR (1990) Better trout habitat: a guide o stream restoration and management. Island Press, Washington, DC, p 320

Hynes HBN (1972) The ecology of running waters. Liverpool University Press, Liverpool, p 555

Johnson DH, Pittman N, Wilder E, Silver JA, Plotnikoff RW, Mason BC, Jones KK, Roger P, O'Neil TA, Barrett C (2001) Inventory and monitoring of salmon habitat in the Pacific Northwest- directory and synthesis of protocols for management/research and volunteers in

Washington, Oregon, Idaho, Montana, and British Columbia. Washington Department of Fish & Wildlife, Olympia, WA, p 212
Jonsson B, L' Abée-Lund JH, Heggbert TJ, Jensen AJ, Johnsen BO, Naesje TF, Saettem M (1991) Longevity, body size, and growth in anadromous brown trout (Salmo trutta). Can J Fish Aquat Sci 48:1838–1945
Leiner S (1995) Biomass and density of brown and rainbow trout in New Mexico streams. Ribarstvo 53:3–24
Lewis SL (1969) Physical factors influencing fish populations in pools of a trout stream. Trans Am Fish Soc 98:14–17
Meehan WR (1991) Influences of forest and rangeland management on salmonid fishes and their habitats. American Fisheries Society, Bethesda. Special Publication 19. 751pp
Meza-Matty IA, Ruiz-Campos G, Daesslé LW, Ruiz-Luna A, López-Lambraño AA, Camarena-Rosales F, Matthews KR (2021) Daily, seasonal, and annual variability of temperature in streams inhabited by the San Pedro Martir trout (*Oncorhynchus mykiss nelsoni*), in Baja California, Mexico, and the predicted temperature for the years 2025 and 2050. J Limnol 80(2):2001. https://doi.org/10.4081/jlimnol.2021.2001
Mitro MG, Lyons JD, Stewart JS (2010) Predicted effects of climate change on the distribution of wild brook trout and brown trout in Wisconsin streams. Págs. 69–76, En: Conserving wild trout (Carline, R.F., C. LoSapio, eds.). Proceedings of the Wild Trout X Symposium. Bozeman, Montana
Murphy ML (1979) Predator assemblages in old-growth and logged sections of small Cascade streams. Thesis, Oregon State University. Corvallis, Oregon. 72pp
O'Connor JE, Chase CG (1989) Uplift of the Sierra San Pedro Mártir Baja California. Tectonics 8: 833–844
Platts WS, McHenry ML (1988) Density and biomass of trout and char in western streams. U.S. For. Serv. Gen. Tech. Rep. INT-241. 17pp
Rinne JW (1982) Movement, home range, and growth of a rare southwestern trout in improved and unimproved habitats. N Am J Fish Manag 2:150–157
Ruiz-Campos G (1989) Repoblación natural por trucha arcoíris (Salmo gairdneri nelsoni) en un transecto del Arroyo San Rafael, Noroeste de la Sierra San Pedro Mártir, Baja California, México. Southwest Nat 34:552–556
Ruiz-Campos G (1993) Bionomía y ecología poblacional de la trucha de la Sierra San Pedro Mártir, Oncorhynchus mykiss nelsoni. Tesis de Doctorado en Ciencias. Universidad Autónoma de Nuevo León, México. 223pp
Ruiz-Campos G (2017) La trucha arcoíris de la Sierra San Pedro Mártir: bionomía, ecología poblacional, hábitat y conservación. Editorial Tirant Lo Blanch, De México, p 280
Ruiz-Campos G, Camarena-Rosales F, González-Acosta AF, Maeda-Martínez AM, De León FJG, Varela-Romero A, Andreu-Soler A (2014) Estatus actual de conservación de seis especies de peces dulceacuícolas de la península de Baja California, México. Rev Mex Biodivers 85:1235–1248
Ruiz-Campos G, Camarena-Rosales F, Varela-Romero A, Sánchez-González S, De La Rosa-Vélez J (2003) Morphometric variation of wild trout populations from Northwestern México (Pisces: Salmonidae). Rev Fish Biol Fish 13:91–110
Ruiz-Campos G, Cota-Serrano P (1992) Ecología alimenticia de la trucha arcoíris (Onchorhynchus mykiss nelsoni) del arroyo San Rafael, Sierra San Pedro Mártir, Baja California, México. Southwest Nat 37:166–177
Ruiz-Campos G, Pister EP (1995) Distribution, habitat, and current status of the San Pedro Mártir rainbow trout, *Oncorhynchus mykiss nelsoni* (Evermann). Bull South Calif Acad Sci 94:131–148
Ruiz-Campos G, Pister EP, Compeán-Jimenez GA (1997) Age and growth of Nelson's trout, *Oncorhynchus mykiss nelsoni*, from Arroyo San Rafael, Sierra San Pedro Mártir, Baja California, México. Southwest Nat 42:74–85

Ruiz-Campos G, Reyes-Valdez CA, Camarena-Rosales F, González-Acosta AF (2016) Relaciones biométricas comparativas de peso y longitud y longitud-longitud entre la trucha dorada mexicana (*Oncorhynchus chrysogaster*) y otras truchas nativas del noroeste de México. Págs. 87–95, En A. Ruiz-Luna y F.J. García De León (eds.). La trucha dorada mexicana. Groppe Impresores, Cd. de México

Ruiz-Campos G, Varela-Romero A (2016) Oncorhynchus mykiss nelsoni (Evermann, 1908). Págs. 221–224, En Los Peces Dulceacuícolas de México en peligro de extinción (G. Ceballos, E. Díaz Pardo, L. Martínez Estévez, y H. Espinosa Pérez, coords.). Fondo de Cultura Económica. de México

Ruiz-Campos G, Villalobos-Ramírez MM (1991) A simple technique for making a streamer-type fish tag. N Am J Fish Manag 11:475–476

Sedell JR, Yuska JE, Speaker RW (1982) Habitats and salmonid distribution in pristine, sediment-rich river valley systems: South Fork Hoh and Queets River, Olympic National Park. Págs. 33-46, En Fish and wildlife relationships in old-growth forest: Proceedings of a symposium (W.R. Meehan, T.R. Merrell, y T.A. Hanley, eds.). American Institute of Fishery Biologists, Morehead City, North Carolina

Sokal RR, Rohlf FJ (2012) Biometry: the principles and practice of statistics in biological research, 4th edn. W. H. Freeman and Co, New York, p 937

Solís-Mendoza M (2016) Caracterización del hábitat de la trucha arcoíris (Oncorhynchus mykiss nelsoni) en la Sierra de San Pedro Mártir, Baja California, y su relación con la densidad y estructura poblacional. Tesis de Maestría en Ciencias, Centro de Investigación Científica y de Educación Superior de Ensenada, Ensenada, Baja California, México. 80pp

Spence BC, Lomnicky GA, Houghes RM, Novitzki RP (1996) An ecosystem approach to salmonid conservation. The ManTech Report 21TR-4501-96-6057. NOAA Fisheries/West Coast Region, Corvallis, OR, p 356

Stewart PA (1970) Physical factors influencing trout density in a small stream. Dissertation, Colorado State University, Fort Collins, CO, 201pp

Surber EW (1937) Rainbow trout and bottom fauna production in one mile of stream. Trans Am Fish Soc 66:193–202

Tamayo JL (1962) Geografía general de México, segunda edición, tomo II: Geografía física. Instituto Mexicano de Investigaciones Económicas, México, pp 249–354

Tamayo JL, West RC (1964) The hydrography of middle America. Págs. 84–121, En Handbook of Middle America. vol. (I.R. Wauchope, ed.). University of Texas Press, Austin

USDA-USFS [United States Department of Agriculture-United States Forest Service] (1990) Pacific Southwest Region habitat typing field guide. Oregon, EUA

Valles-Ríos ME, Ruiz-Campos G (1996) Prevalencia e intensidad de helmintos parásitos del tracto digestivo de la trucha arcoíris *Oncorhynchus mykiss nelsoni* (Pisces: Salmonidae), de Baja California, México. Rev Biol Trop 44-45:579–584

Wesche TA (1980) The WRRI trout cover rating method: development and application. Water Resour. Ser. 78. University of Wyoming, Water Resources Research Institute, Laramie, Wyoming. 46pp

White RJ (1975) Trout population responses to streamflow fluctuation and habitat management in Big Roche-a-Cri Creek, Wisconsin. Verh Int Ver Theor Angew Limnol 19:2469–2477

Mayan Truffles: Notes on the Hypogeous and Subhipogeous Fungi of the Yucatan Peninsula, Mexico

Javier Isaac de la Fuente, Jesús García Jiménez, Gonzalo Guevara Guerrero, León Esteban Ibarra-Garibay, Fortunato Garza-Ocañas, Michael Oswaldo Uitzil-Collí, Juan Pablo Pinzón, and Rafael Peña-Ramírez

Abstract Hypogeous fungi have evolved to produce fruit bodies that develop underground. They have been little studied compared to epigeous macrofungi. Most of the research has been focused on the temperate forests; however, data on these fungi in the tropical zone is still scarce. In this work, 10 genera of hypogeous fungi belonging to the Basidiomycota, Glomeromycota and Mucoromycota phyla from the Yucatan peninsula are discussed. Furthermore, comments on its ecology, distribution and taxonomy are presented.

Keywords Sequestrate fungi · Yucatan · Tropical forests · Mexico

---

J. I. de la Fuente (✉)
Colegio de Postgraduados, Campus Montecillo, Texcoco, Mexico, Mexico

J. G. Jiménez · G. G. Guerrero
Tecnológico Nacional de México, Instituto Tecnológico de Ciudad Victoria, Ciudad Victoria, Tamaulipas, Mexico

L. E. Ibarra-Garibay
Universidad Nacional Autónoma de México, UMDI Juriquilla, Facultad de Ciencias, Laboratorio de Ecología de Artrópodos en Ambientes Extremos, Juriquilla, Querétaro, Mexico

F. Garza-Ocañas
Universidad Autónoma de Nuevo León, Campus Linares, Facultad de Ciencias Forestales, Linares, Nuevo León, Mexico

M. O. Uitzil-Collí
Post Grado en Biociencias, Laboratorio de Micología, Departamento de Botánica, Escuela Nacional de Ciencias Biológicas, Instituto Politécnico Nacional, Ciudad de Mexico, Mexico

J. P. Pinzón
Departamento de Botánica, Facultad de Medicina Veterinaria y Zootecnia, Universidad Autónoma de Yucatán, Mérida, Yucatán, Mexico

R. Peña-Ramírez
Tecnológico Nacional de México, Instituto Tecnológico Superior de Irapuato, Irapuato, Guanajuato, Mexico

F. Garza-Ocañas (ed.), *Sustainable Management of Natural Resources*, Earth and Environmental Sciences Library, https://doi.org/10.1007/978-3-031-33394-1_3

1 Introduction

Hypogeous fungi, also called sequestrate fungi, pseudotruffles, or false truffles, are those fungal species that develop their sporocarps usually underground (Castellano et al. 1986, 2004). These are characterized by a consistent layer (peridium) that encompasses and protects the fertile tissues (gleba) and columella, which is a vestigial stipe. Some authors suggest the use of terms such as pileus, hymenium and stipe for hypogeal species within Agaricales and Boletales.

Taxonomically, most of the sequestrate species can be found within Ascomycota, Basidiomycota and in few species in Glomeromycota and Mucoromycota (Trappe et al. 2009; Castellano et al. 1986; Pegler et al. 1993). Traditionally, hypogeous species were classified as independent taxa but close to their pileate-stipitate counterparts (e.g. *Boletus* and *Gastroboletus*). However, recent studies with molecular evidence have shown that hypogeous and epigeous forms can appear within the same genus (Fig. 1). This has led to many species formerly classified in genera composed solely of sequestrate morphological forms to be recombined and immersed within epigeous genera (Bougher and Lebel 2001; Elliott and Trappe 2018).

Ecologically, many species of hypogeous fungi form mycorrhizal relationships with various species of angiosperm and gymnosperm trees and shrubs which aid in the translocation of nutrients (Trappe et al. 2009; Tedersoo et al. 2010). In addition, many of them are an important part of the diet of mycophagous animals such as mammals and small invertebrates. Economically, some species such as those of the genus *Tuber* are of great commercial importance, since due to their characteristic aroma they are highly appreciated, reaching very high prices in the market (Bonito et al. 2010).

Despite their importance, knowledge of these fungi worldwide is scarce, compared to what is known about epigeous fungi. This is due in large part to the hypogeous habit of these fungi, because in many cases they are not included in the mycobiotic inventories or are not properly sampled, which contributes to little knowledge of them (Sulzbacher et al. 2016). Approximately, more than 150 genera of hypogeous fungi are known worldwide, with North America and Europe being the areas with the greatest amount of data and knowledge about it (Montecchi and Sarasini 2000; Castellano et al. 2004). Like other taxonomic groups of fungi, they have been more studied in temperate forests than in tropical forests. However, in recent years a wide diversity of species has been discovered in the tropical forests of South America, Africa, and Australia. These studies have resulted in the discovery and description of many new species and even new genera (Sulzbacher et al. 2016, 2020; Smith et al. 2015; Castellano et al. 2016; Hosen et al. 2019). In Mexico, most of the research focused on this type of fungi has been limited to northern and central Mexico (Trappe and Guzmán 1971; Gómez-Reyes et al. 2018; Guevara-Guerrero et al. 2008; Guevara et al. 2013; Guevara-Guerrero et al. 2014). Mexico's tropical forests, particularly those found in the Yucatan Peninsula, have been little

Fig. 1 Epigeous and hypogeous forms can occur within the same family or genus. Example: *Cortinarius* (**a**) and *Hymenogaster* (**b**), *Neoboletus* (**c**) and *Xerocomellus* (**d**) and *Lactarius* (**e**) and *Arcangelliela* (now *Lactarius*) (**f**).

explored mycologically and few hypogeous species have been recorded. Although the tropial fungi have been studied since the late nineteenth century, the knowledge of hypogeous fungi is practically nule. It was not until the 80's that the studies of Chío and Guzmán (1982), Guzmán (1982) and Guzmán (1983) documented for the first time in the region some species such as *Redeckera fulva* (Berk. & Broome) C. Walker & A. Schüßler and *Octaviania ciqroensis* Guzmán.

In recent years, some species of hypogeous Agaricales and Phallales have been documented (de la Fuente et al. 2018, 2019, 2020; Uitzil-Collí 2019). These have been found to be potentially associated with some plant communities that have not been well studied mycologically in the region such as the savannah with *Pinus caribaea* and the lowland forests with *Gymnopodium floribundum*, which are well known for harboring an interesting fungal diversity associated with their roots (Tedersoo et al. 2010). This study documents 5 years of mycological research in the Yucatan Peninsula, focused on hypogeous species, presenting data on their distribution, ecological habits, and expectations.

2 Objective of Study

Contribute to the taxonomic, ecological and distribution knowledge of the hypogeous fungi present in the forests of the Yucatan Peninsula.

3 Methodology

The Yucatan Peninsula is placed in southeastern Mexico and comprises the states of Campeche, Quintana Roo, and Yucatan. It does not have high elevations and the highest areas do not exceed 300 meters above sea level. Even if there is not a marked elevation gradient, there is a climatic gradient being drier in the northwest of Yucatan and wetter in southern Campeche and Quintana Roo (Bautista et al. 2012) (Fig. 2). Mycological explorations and collections of the specimens were carried out in various sites of the Yucatan Peninsula since 2010 following the methods of Castellano et al. (1986). The dominant vegetation at these sites corresponds to

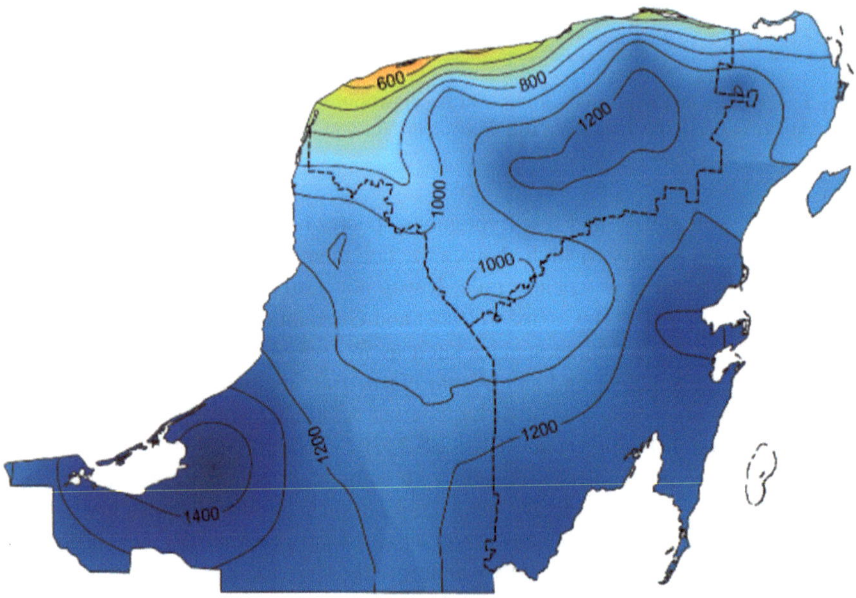

Fig. 2 Precipitation gradient of the Yucatan Peninsula

Fig. 3 Types of vegetation studied in this research. (**a**) Subdeciduous forest. (**b**) Subperennial forest. (**c**) Flooded lowland forest. (**d**) Savannah with pines. (**e**) Secondary vegetation. (**f**) Urban gardens

different common plant communities according to the classification of Valdéz-Hernández and Islebe (2011) which are briefly described below (Fig. 3).

3.1 Subdeciduous Forest

It is characterized by having an arboreal stratum between 10 and 15 m, which in drought season between 50 and 75% of its trees lose their leaves. The characteristic species of this type of vegetation are: *Ceiba pentandra*, *Enterolobium cyclocarpum* and some species of *Ficus*. Other tree species are as follows: *Astronium graveolens, Annona reticulata, Bahuinia divaricata, B. erythrocalix, B. ungulata, Bucida buceras, Bursera simaruba, Caesalpinia gaumeri, C. mollis¸ Cedrela odorata, Cochlospermun vitifolium, Coccoloba acapulcensis, C. barbadensis, C. cozumelensis, C. spicata, Disopyros anisandra, D. campechiana, D. cuneata, D. salicifolia, D. verae- crucis, D. yatesiana, Erythroxylum areolatum, E. confusum, E. rotundifolium, Gliricidia maculata, Guazuma ulmifolia, Gymnopodium floribundum, Gyrocarpus americanus, Havardia albicans, Lonchocarpus guatemalensis, L. longistylus, L. rugosus, L. xuul, L. yucatanensis, Lysiloma latisiliquum, Malmea depressa, Mariosousa dolychostachya, Metopium brownei, Mimosa bahamensis, Oreopanax capitatus, Piscidia piscipula, Pithecellobium dulce, P. lanceolatum, Sapindus saponaria, Sapranthus campechianus, Senegalia gaumeri, Simarouba glauca, Sphinga platyloba, Spondias bombin, Terminalia buceras, Trema micrantha, Vachellia cornigera, V. pennatula, Vitex gaumer*i among other species.

3.2 Sub-Evergreen Forest

It is the vegetation that occupies the largest extension in the Yucatan Peninsula, structurally it is like the high jungle, but they differ in the number of species and the size of the trees, being those of the medium jungle smaller, with heights of between 15 and 25 m. The most representative species of this type of forest are *Swietenia macrophylla, Manilkara zapota, Brosimum alicastrum, Pouteria campechiana, Swartzia cubensis,* and *Sabal yapa*. (Valdéz-Hernández and Islebe 2011).

3.3 Flooded Lowland Forest

This type of vegetation is established in depressions with poorly drained soils. This type of jungle is characteristic of the Yucatan Peninsula and is not found in any other region of Mexico. They are distributed in the form of patches within medium forests. They are communities made up of few species that barely exceed 10 meters in height. The most common species are *Haematoxylum campechianum, Gymnopodium floribundum, Terminalia buceras, Dalbergia glabra, Cameraria latifolia*. Many species of epiphytes (mainly Bromeliaceae and Orchidaceae) are also found (Valdéz-Hernández and Islebe 2011).

3.4 *Pine Savannah*

These areas are home to the only two wild populations of *Pinus caribaea* in Mexico (Macario-Sánchez and Sánchez 2011). They are located at the savannah of Jaguactal, located in the ejido Caobas and in the ejido Pioneros del Río, located in the border area of Quintana. Roo, Campeche and Guatemala. The representative vegetation is composed of low jungle species such as *Gymnopodium floribundum, Erythroxylum confusum, Cameraria latifolia* and savannah species such as *Curatella americana, Byrsonima crasifolia* and *Crescentia cujete* (Macario-Sánchez and Sánchez 2011).

3.5 *Secondary Vegetation/Urban Gardens*

This type of vegetation contains species typical of secondary succession such as *Leucaena leucocephala, Metopium brownei, Lysiloma latisiliquum, Coccoloba spicata* and some introduced as *Delonix regia* and *Casuarina equisetifolia*. The secondary vegetation depends a lot on which area of the peninsula it is, the type of soil and the vegetation that existed before so it is very complicated to characterize it.

The collection of the specimens was carried out following the criteria of Castellano et al. (1986, 2004). The search for the fruiting bodies was carried out under the leaf litter and in the upper layer of the substrate at a depth of no more than 15 cm (Gómez-Reyes et al. 2012). Cuts were made in the dehydrated specimens to observe and measure the structures of taxonomic importance. The specimens are deposited in the mycological herbarium "José Castillo Tovar" at the Technological Institute of Ciudad Victoria (ITCV) and the herbarium "Alfredo Barrera Marín" of the Autonomous University of Yucatan (UADY).

4 Results

More than 70 collections grouped into three phyla (Basidiomycota, Glomeromycota and Mucoromycota), 10 families and 23 taxa were studied. The most representative division was Basidiomycota with 19 taxa while Mucoromycota presented three taxa and Gloeromycota with one taxon; It is worth mentioning that in this study no hypogeous ascomycetes were located. The most abundant species was *Redeckera fulva* (Glomeroycota) which is very common in the tropical forest of the peninsula (Guzmán 1983, 2003). The state of Quintana Roo presented the greatest diversity of species (14) followed by Campeche (9) and Yucatan (6). The sub-evergreen forest and the flooded lowland forest presented the greatest diversity of species richness (eight species) while the subdeciduous forest presented five species; the secondary vegetation and the savannah with pines presented the least rich diversity when only five and two species were registered respectively. Photographs of the species

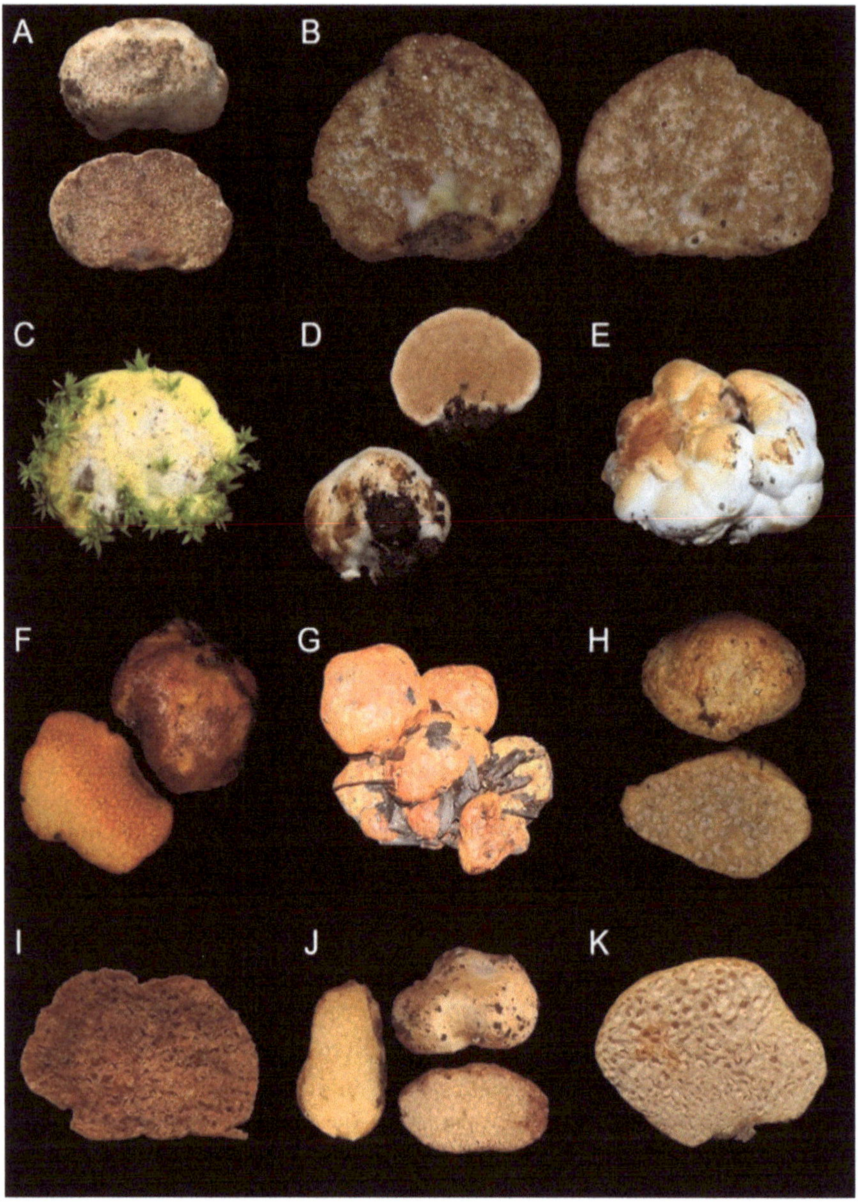

Fig. 4 Hypogeous fungi from the Yucatan Peninsula. (**a–d**) *Endogone* spp. (**e**) *Redeckera fulva* (**f**) *Mayamontana coccolobae* (**g**) *Mayamontana* sp., (**h**) *Stephanospora mayana* (**i**) *Stephanospora xibalba* (**j**, **k**) *Stephanospora* spp.

discussed are shown in Figs. 4 and 5. The species are ordered taxonomically according to Tedersoo et al. (2018) and Wijayawardene et al. (2020).

Fig. 5 Hypogeous fungi from the Yucatan Peninsula. (**a**) *Melanogaster coccolobae*. (**b**, **c**) *Scleroderma* spp. (**d**, **e**) *Sclerogaster* spp. (**f**–**i**) *Restingomyces* spp. (**j**) *Lactarius* sp.

4.1 Phylum Glomeromycota

Order Diversisporales, Family Diversisporaceae
Redeckera C. Walker & A. Schüßler

The only macroscopic species capable of forming subhypogeal sporocarps of this division recorded in the Yucatan Peninsula is *R. fulva* (as *Glomus fulvum*). This species is characterized by the subhypogeous, globose or subglobose sporome, whitish and smooth, which quickly stains yellowish when touched. The gleba is soft in texture, without locules. Spores are variable in size, usually between 50 and 100 μm long.

They are common fungi in the Yucatan Peninsula and can be found in secondary vegetation, deciduous forests, lowlands and savannahs.

This species is very common, and its colors make it easy to differentiate from other species. It has been recorded in Yucatan by Guzmán (1983, 2003), Chay-Casanova and Medel (2000), and Pompa-González et al. (2011).

4.2 Phylum Mucoromycota

Order Endogonales

Family Endogonaceae

Endogone Link

The species of this genus show globose or sub globose or irregular sporocarps, smooth or cottony, yellowish or whitish, with interwoven hyphae covering the sporangia, randomly occuring within the sporocarpst. The spores are usually large (more than 50 μm) with ornamentations or with a mantle of hyphae attached.

Species of the genus can be found in medium sub-evergreen forests, sometimes associated with moss species.

Some species are similar in morphology to *Redeckera fulva*, although the latter does not present sporangia. Some species found in Quintana Roo are similar to *Endonogone lactiflua* and *E. pisciformis*, however, these species are associated with plants from temperate forests (Pegler et al. 1993; Trappe et al. 2009; Blaszkowski et al. 2004). Therefore, it is possible that they are different species, so the corresponding taxonomic studies (morphological and molecular) will continue to elucidate their identity.

4.3 Phylum Basidiomycota, Order Agaricales

Family Stephanosporaceae

Mayamontana Castellano, Trappe & Lodge

The only species of the genus is characterized by the reddish basidioma, the thin, almost transparent peridium that allows to see the locules. The hymenium is composed of rounded or angular locules, reddish or orange in color. It presents globose to ellipsoid basidiospores, smooth or rough, with conspicuous hilar appendage.

It usually grows in flooded lowland forest under *Coccoloba, Gymnopodium,* and *Metopium*. It has also been found in urban gardens growing under *Lysiloma latisiliquum*.

Mayamontana coccolobae was described in Belize (Castellano et al. 2007) and has been reported in the Yucatan Peninsula in the states of Campeche and Quintana Roo (de la Fuente et al. 2018). It is easy to differentiate in field due to the coloration of the basidiomata. A second species has been found in Quintana Roo which has flattened, caespitose basidiomata and larger spores than those reported for *M. coccolobae*. Because of this, it is likely that two different species may occur in Mexico, so it is suggested to continue with morphological and molecular studies.

4.4 Division Basidiomycota, Order Agaricales

Family Stephanosporaceae

Stephanospora Pat.

The fungi of this genus are characterized by their hypogeous or subhypogeous, globose or sub globose basidiomata, with orange, reddish or yellowish colors. The hymenium is composed of rounded or angular locules, pale yellow, grayish, yellowish or orange. Microscopically they are distinguished by their ellipsoid or globose basidiospores, ornamented with spines, forming a "crown" at the base of the basidiospore.

They are common in lowland forests and to a lesser extent in subdeciduous forest forests and secondary vegetation of the Yucatan Peninsula. They grow under *Gymnopodium, Coccoloba,* and *Lysiloma*.

Two species of *Stephanospora* have been described in recent years: *Stephanospora mayana* and *Stephanospora xibalba* (de la Fuente et al. 2019, 2020) and three other species remain undescribed. Although most species of *Stephanospora* have a well-developed crown, some species of the peninsula have a very superficial one as in S. *xibalba*.

4.5 Phylum Basidiomycota, Order Boletales

Family Boletaceae

Octaviania Vittad.

Species with hypogeous basidiomata, usually globose, cream, or yellowish colors, with loculated and gelatinous gleba. Microscopically they are characterized by globose basidiospores ornamented with large cones of more than 2 μm.

They are generally associated with pines and oaks, however, in the Yucatan Peninsula they are possibly associated with Polygonaceae. *Octaviania ciqroensis* was described from Quintana Roo, but no data are available on the possible host (Guzmán 1982). It is a rare species since only the type collection is known. The basidiospores have very small spines compared to other species of the genus and a dry gleba, characteristics that resemble *Stephanospora* so a taxonomic and molecular review of the species is recommended.

4.6 Phylum Basidiomycota, Order Boletales

Family Paxillaceae

Melanogaster Corda

They present hypogeous or subhypogeous basidiomata, globose, subglobose, usually brown or orange peridium, gleba with locules filled with a gelatinous substance of black color. Some species have a smell ranging from sweet and fruity to garlic-like odors. They have dark brown ellipsoid basidiospores. Usially they grow under pines, oaks and some tropical trees such as *Coccoloba*.

Records of *Melanogaster* species are still scarce, only three species are known from Mexico associated with pine and oak forests (Trappe & Guzmán 1971; Cázares et al. 1992, 2008). Séne et al. (2012) report a species of *Melanogaster* from the roots of *Coccoloba uvifera* in the Caribbean islands, however, due to the size of the specimen (200 μm) it was not described. So far, Only *Melanogaster coccolobae* is found in the tropical forests of Mexico.

4.7 Phylum Basidiomycota, Order Boletales

Family Sclerodermataceae

Scleroderma Pers.

Basidiomata rarely hypogeous, usually epigeous. They have a thick, usually dry, scaly, or cracked peridium that opens to expose the dusty, brownish gleba. Basidiospores are globose, spiny to reticulated.

In the Yucatan Peninsula there are five species of *Scleroderma* which are associated with *Coccoloba diversifolia, C. spicata, C. uvifera* and *Gymnopodium floribundum.*

Scleroderma albidum Pat. & Trab, *S. areolatum* Ehrenb., *S. bermudense* Coker, *S. nitidum* Berk., and *S. siniamariense* Mont. are epigeous species and have been described by Guzmán et al. (2013). However, in recent years, a morphologically distinct species has been discovered with lilac and thick peridium and basidiospores with short spines. Morphologically it could be confused with some species of *Elaphomyces,* however, the specimens do not present asci. A second species of lilac peridium but much thicker (up to 7 mm) and gleba with irregular chambers, characteristics that are not typical of the genus. These species are tentatively accommodated in *Scleroderma* although it could represent a different lineage.

4.8 Phylum Basidiomycota, Order Boletales

Family Sclerogastraceae

Sclerogaster R. Hesse

Hypogeousl basidiomata, often gregarious, whitish, some staining when touched. They have greenish or yellowish gleba of cartilaginous appearance, without columella or poorly developed. Microscopically, they are characterized by globose basidiospores ornamented with small spines or small cones. In the Yucatan Peninsula they are found in association with *Pinus caribaea* and in lowland forest with *Coccoloba diversifolia*. Species of the genus are rare in the peninsula and only few basidiomata have been collected. According to Trappe et al. (2009) it is a little-known genus that needs taxonomic revision.

4.9 Phylum Basidiomycota, Order Phallales

Family Trappeaceae

Restingomyces Sulzbacher, Grebenc & Baseia

Hypogeous to subhipogeous basidiomata, with cottony and whitish peridium, rarely dry and somewhat cracked, with greenish or brownish gleba, some with sterile base and dendriform columella. Basidiospores are ellipsoid, smooth, or reticulate. Several specimens have only been collected in Campeche and Yucatan under *Gymnopodium floribundum* and *Manilkara sapota*.

Macromorphologically they are similar to *Hysterangium* by the whitish peridium, the gelatinous or cartilaginous gleba and the columella, however, micro morphologically they can be separated by the shape of the basidiospore, which is fusoid in *Hysterangium* in addition to these can have a detachable peridium, a rough or smooth utricle and a different phylogenetic position (Guevara-Guerrero et al. 2008; Hosaka et al. 2006; Sulzbacher et al. 2016).

4.10 Phylum Basidiomycota, Order Russulales

Family Russulaceae

Lactarius (Pers.) Roussel

Species of this genus typically have epigeous basidiomata, although hypogeous species can also be found. They are characterized by the presence of latex when cut, spherocysts in the trama and amyloid basidiosspores with warts or reticulum.

In the Yucatan Peninsula, it has only been found growing in lowland forest with *Gymnopodium floribundum*. They are difficult to observe because of their small size (less than one cm.). The specimens of the peninsula have a whitish or pale-yellow color, with a depression at the base, with irregular orange-brown locules, exuding a scarce, transparent latex. The characteristics of the specimen coincide well with the concept of *Arcangelliela* (Trappe et al. 2009) however, like other genera of hypogeous fungi within the Russulaceae family such as *Zelleromyces, Martelia, Cystangium, Macowanites,* have been recombined and placed within agaricoid genera such as *Russula* and *Lactarius* (Verbeken and Nuytinck 2010; Elliott and Trappe 2018).

5 Discussion

The diversity of hypogeous fungi is low when compared to other groups of fungi well studied in the tropical forests of the Yucatan Peninsula such as the Agaricales, Xylariales and Polyporales (Guzmán 2003; Pompa-González et al. 2011). However, this is largely due to the lack of specialized mycologists in this group.

Recently, research carried out by the authors has resulted in several new species and a new records from Mexico (de la Fuente et al. 2018, 2019, 2020) and informative studies are carried out on this little-known group in the peninsula (Uitzil-Collí 2019). It is expected that, with increased sampling effort and more detailed taxonomic studies, the number of species reported will increase and as well as the discovery of species new to science.

Another factor affecting the diversity of hypogeous fungi is the lack of host trees. However, in the Yucatan Peninsula it is possible to find tree species such as *Coccoloba*, *Gymnopodium* and *Pinus*, which have been documented as ectomycorrhizal hosts of many species of fungi in the Caribbean (Ortíz-Santana et al. 2007; Tedersoo et al. 2010, Põlme et al. 2017; Singer et al. 1983, Bandala and Montoya 2015) but scarcely of hypogeous fungi. Although the genus *Pinus* is well known as one of the most common hosts of ectomycorrhizal fungi, fungal diversity in tropical pine savannas in the Mexico is scarce. Possibly, since the savannah is flooded in the rainy season, it does not allow a correct development of fungi. Further, more mycological explorations are needed to document the mycorrhizal fungi from this interesting community.

Although sequestrate species diversity is not very high in Yucatan, it is very likely that several specimens found in this study represent species new to science. Some of the specimens studied here were found in interesting plant communities, for example, the lowland forest with *Gymnopodium floribundum* harbors a great diversity of ectomycorrhizal fungi and that has been scarcely studied in Mexico (Bandala and Montoya 2015). It is interesting to observe the diversity of sympatric species of Stephanosporaceae which can be found in lowland, deciduous forests, and even secondary vegetation. As in the rainforests of Guyana (Sulzbacher et al. 2016), most hypogeous fungi belong to Basidiomycota and no species of hypogeous ascomycetes have been found in the Yucatan Peninsula so far.

Ecological studies have not been carried out on mycophagy by native wildlife, however, it is interesting to mention that some species of *Stephanospora* emit aromas with some resemblance to those of small regional fruits, which could facilitate their dispersion by attracting animals that consume them. Other species of *Melanogaster, Endogone*, and *Lactarius* have interesting aromas in order to attract animals to disperse their spores. It is common to observe sporocarps of *Redeckera fulva* with signs of bites, possibly of small invertebrates. Little has been studied about the consumption of mushroom sporocarps by native fauna, so it is recommended to continue with the study of these ecological relationships.

6 Conclusion

The Yucatan Peninsula is home to an interesting diversity of hypogeous species, distributed in the phyla Basidiomycota and Glomeromycota, within 10 families and 23 taxa. The state of Quintana Roo harbors 14 taxa, followed by Campeche with 9 and finally Yucatan with 6. Most of the sampling was carried out in deciduous and lowland forest, however, there are still areas that need a greater sampling effort. For the above, it is suggested to continue with taxonomic, phylogenetic, and ecological studies, among others, that allow us to better understand the dynamics and relevance that this type of fungi play in the tropical ecosystems of the region.

References

Bandala VM, Montoya L (2015) *Gymnopodium floribundum* trees (Polygonaceae) harbour a diverse ectomycorrhizal fungal community in the tropical deciduous forest of southeastern Mexico. Res Rev 4:73–75

Bautista F, Palacio G, Páez-Bistraín R, Carmona-Jiménez ME, Delgado-Carranza C, Cantharell W, Tello H (2012) Geografía de suelos regional: Península de Yucatán. In: Krasilnikov P, Jiménez-Nava FJ, Reyna-Trujillo T, García-Calderón NE (eds) Geografía de suelos de México. Prensas de ciencias, Mexico City, pp 335–402

Blaszkowski J, Adamska I, Czerniawska B (2004) *Endogone lactiflua* (Zygomycota, Endogonales) ocurrs in Poland. Acta Soc Bot Pol 73:65–69

Bonito GM, Gryganskyi AP, Trappe JM, Vilgalys R (2010) A global meta-analysis of *Tuber* ITS rDNA sequences: species diversity, host associations and long-distance dispersal. Mol Ecol 19(22):4994–5008

Bougher NL, Lebel T (2001) Sequestrate (truffle-like) fungi from Australia and New Zealand. Aust Syst Bot 14:439–484

Castellano MA, Elliot TF, Truong C, Séné O, Dentinger BTM, Henkel TW (2016) *Kombocles bakaiana* gen. sp. nov. (Boletaceae), a new sequestrate fungus from Cameroon. IMA Fungus 7: 239–245

Castellano MA, Trappe JM, Lodge DJ (2007) *Mayamontana coccolobae* (Basidiomycota), a new secuestrate taxon from Belize. Mycotaxon 100:289–294

Castellano MA, Trappe JM, Luoma DL (2004) Sequestrate fungi. In: Muller G, Bills G, Foster M (eds) Biodiversity of fungi inventory and monitoring methods. Elsevier Academic Press, Burlington, pp 197–213

Castellano MA, Trappe JM, Maser Z, Maser S (1986) Key to spores of the genera of hypogeous fungi of North America, with reference to animal mycophagy. Mad River Press, Eureka, p 185

Cázares E, García J, Castillo J, Trappe JM (1992) hypogeous fungi from northern Mexico. Mycologia 84:341–359

Cázares E, Trappe JM, Guevara G, García J (2008) *Melanogaster mynisporus* sp. nov. A new sequestrate member of the Boletales from Mexico. Rev Mex Mic 28:67–69

Chay-Casanova JA, Medel R (2000) Hongos citados para el jardín botánico y áreas adyacentes. In: Sánchez O, Islebe G (eds) El Jardín Botánico Dr. Alfredo Barrera Marín. Fundamentos y estudios particulares. Conabio-Ecosur, Mexico City, pp 115–124

Chío RE, Guzmán G (1982) Los hongos de la península de Yucatán I las especies de macromicetos conocidas. Biotica 7:385–398

Elliott TF, Trappe JM (2018) A worldwide nomenclature revision of sequestrate *Russula* species. Fung Syst Evol 1:229–242

de la Fuente JI, Guevara-Guerrero G, López CY, García-Jiménez J (2018) First record of *Maymontana coccolobae* (Stephanosporaceae: Agaricales) from Mexico. Stud Fungi 3:34–38

de la Fuente JI, Guevara-Guerrero G, Oros-Ortega I, Sánchez-Zavalegui R, Córdoba-Lara I, García-Jiménez J (2019) *Stephanospora mayana* (Stephanosporaceae, Russulales), a new sequestrate fungus from Yucatán Peninsula, Mexico. Mycokeys 48:115–124

de la Fuente JI, Pinzón JP, Guzmán-Dávalos L, Uitzil-Colli MO, Bahram M, Lebel T (2020) A new *Stephanospora* (Agaricales, Basidiomycota) from the Yucatan peninsula, Mexico. Phytotaxa 436:63–71

Gómez-Reyes VM, Hernández-Salmerón IR, Terrón-Alfonso A, Guevara-Guerrero G (2012) Estudio taxonómico de *Elaphomyces* spp. (Ascomycota, Eurotiales, Elaphomycetaceae) de Michoacán, México. Rev Mex Mic 36:57–62

Gómez-Reyes VM, Vázquez-Marrufo G, Ortega-Gómez AM, Guevara-Guerrero G (2018) Ascomicetos hipogeos de la región occidental del Sistema Volcánico Transversal, México. Acta Bot Mex 125:37–48

Guevara G, Bonito G, Cázares E (2013) Revisión del género *Tuber* (Tuberaceae: Pezizales) de Mexico. Rev Mex Biodivers 84:39–49

Guevara-Guerrero G, Castellano MA, García-Jiménez J, Cázares-González E, Trappe JM (2008) *Hysterangium* (Hysterangiales, Hysterangiaceae) from northern Mexico. Rev Mex Mic 28:95–100

Guevara-Guerrero G, Cázares-González E, Bonito G, Healy RA, Stielow B, García J, Garza-Ocañas F, Castellano MA, Trappe JM (2014) Hongos Hipógeos de Tamaulipas. In: Correa SA, Horta J, García J, Barrientos L (eds) Biodiversidad Tamaulipeca. Vol. 2. Tecnológico Nacional de México-Instituto Tecnológico de Ciudad Victoria, México City, pp 87–102

Guzmán G (1982) New species of fungi from Yucatán Peninsula. Mycotaxon 16:249–261

Guzmán G (1983) Los hongos de la península de Yucatán II. Nuevas exploraciones y adiciones micológicas. Biotica 8:71–78

Guzmán G (2003) Los hongos de El Edén Quintana Roo (introducción a la micobiota tropical de México). INECOL-CONABIO, Xalapa, p 316

Guzmán G, Cortés-Pérez A, Guzmán-Dávalos L, Ramírez-Guillén F, Sánchez-Jácome MR (2013) An emendation of *Scleroderma*, new records, and review of the known species in Mexico. Rev Mex Biodivers 84:173–191

Hosaka K, Bates ST, Beever RE, Castellano MA, Colgan W III, Domínguez LS, Nouhra ER, Geml J, Giachini AJ, Kenney SR, Simpson NB, Spatafora JW, Trappe JM (2006) Molecular phylogenetics of the gomphoid-phalloid fungi with an establishment of the new subclass Phallomycetidae and two new orders. Mycologia 98(6):949–959

Hosen MI, Zhong X-J, Gates G, Orihara T, Li T-H (2019) Type studies of *Rossbeevera bispora*, and a new species of *Rossbeevera* from south China. MycoKeys 51:1–14

Macario-Sánchez PA, Sánchez LC (2011) Pino tropical. In: Pozo C, Armijo N, Calmé S (eds) Riqueza biológica de Quintana Roo. Gobierno del Estado de Quintana Roo-Programa de pequeñas donaciones, Mexico City, pp 52–55

Montecchi A, Sarasini M (2000) Funghi ipogei d'Europa. Associazione Micologica Bresadola, Fondazione Centro Studi Micologici, 1–714

Ortíz-Santana B, Lodge DJ, Baroni TJ, Both EE (2007) Boletes from Belize and Dominican republic. Fungal Divers 27:247–416

Pegler DN, Spooner BM, Young TWK (1993) British truffles. A revision of British hypogeous fungi. The Board of Trustees of the Royal Botanical Gardens, Kew, 215pp

Põlme S, Bahram M, Kõljalg U, Tedersoo L (2017) Biogeography and specificity of ectomycorrhizal fungi of *Coccoloba uvifera*. In: Tedersoo L (ed) Biogeography of mycorrhizal symbiosis, University of Tartu, pp 345–359

Pompa-González A, Aguirre-Acosta CE, Encalada-Oliva AV, De Anda-Jáuregui A, Cifuentes-Blanco J, Valenzuela-Garza R (2011) Los macromicetos del jardín botánico de ECOSUR "Dr. Alfredo Barrera Marín" Puerto Morelos, Quintana Roo. Comisión Nacional para el Conocimiento y Uso de la Biodiversidad, México City, 108pp

Séne S, Avril R, Chaintreuil C, Geoffroy A, Ndiaye C, Diédhiou AG, Sadio O, Courtecuisse R, Sylla SN, Selosse MA, Bâ A (2012) Ectomycorrhizal fungal communities of *Coccoloba uvifera* (L.) L. mature trees and seedlings in the neotropical coastal forests of Guadeloupe (Lesser Antilles). Mycorrhiza 25:547–549

Singer R, Araujo I, Ivory MH (1983) Ectotrophically mycorrizal fungi of the neotropical lowlands, especially central Amazonia. Beihefte zur Nova Hedwigia 77:7–335

Smith ME, Amses KR, Elliott TF, Obase K, Aime MC, Henkel TW (2015) New sequestrate fungi from Guyana: *Jimtrappea guyanensis* gen. sp. nov., *Castellanea pakaraimophila* gen. sp. nov., and Costatisporus cyanescens gen. sp. nov. (Boletaceae, Boletales). IMA Fungus 6:297–317

Sulzbacher MA, Grebenc T, Giachini AJ, Baseia IG, Nouhra ER (2016) Hypogeous sequestrate fungi in South America – how well do we know them? Symbiosis. https://doi.org/10.1007/s13199-016-0461-4

Sulzbacher MA, Orihara T, Grebenc T, Wartchow F, Smith ME, Martín MP, Giachini AJ, Baseia IG (2020) *Longistriata flava* (Boletaceae, Basidiomycota) – a new monotypic sequestrate genus and species from Brazilian Atlantic Forest. MycoKeys 62:53–73

Tedersoo L, May TW, Smith ME (2010) Ectomycorrhizal lifestyle in fungi: global diversity, distribution, and evolution of phylogenetic lineages. Mycorrhiza 20:217–263

Tedersoo L, Sánchez-Ramírez S, Koljalg U, Bahram M, Döring M, Schigel D, May T (2018) High-level classification of the Fungi and a tool for evolutionary ecological analyses. Fungal Divers 90:135–159

Trappe JM, Guzmán G (1971) Notes on some hypogeous fungi from Mexico. Mycologia 63:327–332

Trappe JM, Molina R, Luoma DL, Cázares E, Pilz D, Smith JE, Castellano MA, Miller SL, Trappe MJ., (2009) Diversity, ecology, and conservation of truffle fungi in forests of the Pacific Northwest. General Technical Report PNW-GTR-772, Department of Agriculture, Forest Service, Pacific Northwest Research Station Portland, Oregon, 194pp

Uitzil-Collí MO (2019) Hongos hipógeos: Tesoros subterráneos del trópico mexicano. Blog de la revista de biología tropical. Universidad de Costa Rica

Valdéz-Hernández M, Islebe GA (2011) Tipos de vegetación en Quintana Roo. In: Pozo C, Armijo N, Calmé S (eds) Riqueza biológica de Quintana Roo. Gobierno del Estado de Quintana Roo-Programa de pequeñas donaciones, Mexico City, pp 32–36

Verbeken A, Nuytinck J (2010) Not every milkcap is a *Lactarius*. Scr Bot Belg 51:162–168

Wijayawardene NN, Hyde KD, Al-Ani LKT, Tedersoo L, Haelewaters D, Rajeshkumar KC, Zhao RL, Aptroot A, Leontyev DV, Saxena RK, Tokarev YS, Dai DQ, Letcher PM, Stephenson SL, Ertz D, Lumbsch HT, Kukwa M, Issi IV, Madrid H, Phillips AJL, Selbmann L, Pfliegler WP, Horváth E, Bensch K, Kirk PM, Kolaříková K, Raja HA, Radek R, Papp V, Dima B, Ma J, Malosso E, Takamatsu S, Rambold G, Gannibal PB, Triebel D, Gautam AK, Avasthi S, Suetrong S, Timdal E, Fryar SC, Delgado G, Réblová M, Doilom M, Dolatabadi S, Pawłowska J, Humber RA, Kodsueb R, Sánchez-Castro I, Goto BT, Silva DKA, de Souza FA, Oehl F, da Silva GA, Silva IR, Błaszkowski J, Jobim K, Maia LC, Barbosa FR, Fiuza PO, Divakar PK, Shenoy BD, Castañeda-Ruiz RF, Somrithipol S, Lateef AA, Karunarathna SC, Tibpromma S, Mortimer PE, Wanasinghe DN, Phookamsak R, Xu J, Wang Y, Tian F,

Alvarado P, Li DW, Kušan I, Matočec N, Maharachchikumbura SSN, Papizadeh M, Heredia G, Wartchow F, Bakhshi M, Boehm E, Youssef N, Hustad VP, Lawrey JD, Santiago ALCMA, Bezerra JDP, Souza-Motta CM, Firmino AL, Tian Q, Houbraken J, Hongsanan S, Tanaka K, Dissanayake AJ, Monteiro JS, Grossart HP, Suija A, Weerakoon G, Etayo J, Tsurykau A, Vázquez V, Mungai P, Damm U, Li QR, Zhang H, Boonmee S, Lu YZ, Becerra AG, Kendrick B, Brearley FQ, Motiejūnaitė J, Sharma B, Khare R, Gaikwad S, Wijesundara DSA, Tang LZ, He MQ, Flakus A, Rodriguez-Flakus P, Zhurbenko MP, McKenzie EHC, Stadler M, Bhat DJ, Liu JK, Raza M, Jeewon R, Nassonova ES, Prieto M, Jayalal RGU, Erdoğdu M, Yurkov A, Schnittler M, Shchepin ON, Novozhilov YK, Silva-Filho AGS, Liu P, Cavender JC, Kang Y, Mohammad S, Zhang LF, Xu RF, Li YM, Dayarathne MC, Ekanayaka AH, Wen TC, Deng CY, Pereira OL, Navathe S, Hawksworth DL, Fan XL, Dissanayake LS, Kuhnert E, Grossart HP, Thines M (2020) Outline of fungi and fungus- like taxa. Mycosphere 11(1): 1060–1456

Bioenergetic Potential of the Huizache *Vachellia farnesiana* (L.) Willd

Sandy Juliana Hernández-Mata, Fortunato Garza-Ocañas, Horacio Villalón-Mendoza, José Rodolfo Goche-Telles, and Artemio Carrillo-Parra

Abstract The intensive use of thornscrubs in northeastern Mexico, coupled with environmental factors has allowed for the increase of vegetation cover and this is dominated by shrub species such as *Vachellia farnesiana* (L.) Willd. Firewood and/or charcoal are an alternative for the sustainable use of these species. To achieve this, it is necessary to establish growth and energy parameters for the species. In this research work, the volume of firewood produced, and the characteristics of firewood and charcoal of the species were quantified. The study area is located at the CIPA (Agricultural Production Research Center of the Universidad Autónoma de Nuevo León). Forest production was determined through forest inventories on plots of 4 m × 25 m, the quality of biofuel was determined by proximal analyses of firewood and charcoal made under laboratory conditions. The results were analyzed as a 2 × 4 factorial design, with the factors being site (2) and diametric category (4). Comparisons were made with analysis of variance with significance values of $p < 0.05$. The number of trees per diametric category, the volume of firewood, the values of the proximate analyses, as well as the calorific value, allow to consider V. *farnesiana* as an alternative species to produce energy if it is combined in a comprehensive use with ecological criteria in the grassland areas of northeastern Mexico.

Keywords Biofuels · Native species · Production

S. J. Hernández-Mata · F. Garza-Ocañas · H. Villalón-Mendoza
Universidad Autónoma de Nuevo León (UANL), Facultad de Ciencias Forestales, Linares, Nuevo León, Mexico

J. R. Goche-Telles
Universidad Juárez del Estado de Durango, Facultad de Ciencias Forestales, Durango, Durango, Mexico

A. Carrillo-Parra (✉)
Universidad Juárez del Estado de Durango, Instituto de Silvicultura e Industria de la Madera, Ciudad Universitaria, Durango, Durango, Mexico
e-mail: acarrilloparra@ujed.mx

F. Garza-Ocañas (ed.), *Sustainable Management of Natural Resources*, Earth and Environmental Sciences Library, https://doi.org/10.1007/978-3-031-33394-1_4

1 Introduction

The Tamaulipas thorscrub (Tt) has a long history of silvo-agricultural use, has played a relevant role in the economy of northeastern Mexico since the late sixteenth century. The area occupied by the MET covers an approximate 200,000 km^2 of northeastern Mexico and southern Texas (Ledezma et al. 2018), however, the area is reduced at an average annual rate of 600 km^2. This could lead to a series of negative effects on vegetation, such as loss of species and genetic material, the interruption in pollination and dispersal of seeds, as well as the invasion of generalist species (Molina-Guerra 2013). The fragmentation of ecosystems represents one of the main causes of biodiversity loss at the global level (Molina-Guerra 2013).

Among the arvenses species of the scrub stands out *Vachellia farnesiana* (L.) Willd. (Cavalcante and Cox 2016), this is developed on the banks of roads, streams, in abandoned plots, land with disturbance, successional land (acahuales) and rural sites. It is found where warm climates (Aw) and semi-warm A(C) predominate, in regions that have up to 900 mm of annual precipitation and temperatures that vary from 5 to 30 °C.

It thrives in a wide variety of soils from very clayey to very sandy such as rendzine, xegorendzine, vertisol, sandy, moist, limestone, gypsum and alluvial soils.

Vachellia farnesiana is a secondary species and an important element of the vegetation that occurs in the tropical deciduous forest. It forms dense associations called "huizachales" and is indicative of disturbed sites. The huizache has the potential to occupy a wider range of distribution than the current one and is considered an arvense species and could be an important source of biomass for energy generation.

Biofuels are considered as an alternative to replace fossil fuels, which will reduce the effects caused by them (Callejas and Quezada 2009; Collard and Blin 2014). In 2017, biofuels supplied 14.9% of global energy demand (IEA 2018).

The main objective of this work is to determine the energy feasibility of biomass (firewood) generated by the proliferation of trees of the species *Vachellia farnesiana* in cattle grazing areas of northeastern Mexico and that meets parameters of integral use, generator of solid biofuel with the minimum ecological impact.

2 Methodology

2.1 Study Area

The study was carried out at the Agricultural Production Center of the Autonomous University of Nuevo León at the geographical coordinates 24° 47' 37.88N/ 99° 32' 26.19O. The altitude of the area is 376 m. The climate according to the Köppen classification is (A) C (x') (w'') to (e) corresponding to an extreme or semi-warm subhumid climate basically due to the geographical position in the subtropical zone of high pressures, with influence of wet trade winds and cold winds from the north during the winter (Hernández and Ybarra 2008).

2.2 Sampling

The study was conducted in a grazing area where V. *farnesiana* trees have developed freely for 20 years. The area was subdivided into three sites, at each site plots of 2 m × 50 m (100 m^2) were established, in each site three trees were randomly selected from each of the diametric category 5, 10, 15 and 20 cm. The trees were felled and cut into small sections, the logs (> 0.05 m in diameter) and the branches (< 0.05 m) were separated. Figure 1 shows the process of felling and chopping trees.

2.3 Quantification of Firewood

The volume of firewood was determined from the diametric categories 5, 10, 15, and 20 cm. All the logs and branches were weighed and measured in green condition. The first log (1 m in length) and a section of approximately 10% of the branches were collected and taken to the laboratory for further analysis. Figure 2 shows the process used to weigh firewood from felled trees on each plot.

Fig. 1 Felling and cutting tree of *Vachellia farnesiana* to determine biomass

Fig. 2 (**a**) Weighing process and measurement of a branch (**b**) Weight of fine branches of *Vachellia farnesiana*

2.4 Volume of Firewood and Biomass

The logs obtained from each of the three downed *V. farnesiana* trees of each diametric category were sawn to make wooden cubes of 2 cm per side. Four cubes per each diametric category and each tree were placed on a stove at 105 ± 3 °C for a period of 24 h or until the weight of the cubes remained constant. The cubes were weighed, and the dimensions of the tangential, radial and longitudinal side were measured in green and in anhydrous condition to determine the volume.

A procedure like the previous one was used to determine the moisture content of the branches, just modifying the time and the used temperature was 60 °C, and the branches were separated according to size into very thin (< 5 mm), thin (5–10 mm), thick (> 10 mm), and leaves.

2.5 Charcoal Yield and Charcoal Quality

The charcoal of *V. farnesiana* was elaborated in laboratory conditions from cubes of 2 cm on each side. The cubes were placed in iron tubes in a muffle at 750 °C for 30 min according to the procedure established by Santos et al. (2012). The pyrolysed cubes were removed from the tubes, measured and their mass determined. Figure 3 shows the procedure for the elaboration of the charcoal and the preparation of the samples to perform the proximal analyses. The charcoal yield was determined according to the equation established by Vogel and Wolf (2012).

$$\text{Charcoal yield} = \text{Cho}/\text{Fwo}$$

Where: Po is the mass of anhydrous charcoal and Fwo is the mass of anhydrous firewood.

Fig. 3 (**a**) Placement of the firewood samples in the iron tubes, (**b**) Elaboration of the charcoal in the muffle at 750 °C, (**c**) Grinding of charcoal and (**d**) Screening of charcoal

The buckets were ground, sieved in a mesh of the number 40, the ground material was analyzed to determine the moisture content (%), volatile material (%), ash (%) and fixed carbon (%) (proximal analysis) according to the procedure detailed below:

Moisture from the crucibles was removed, for this they were placed in the muffle at a temperature of 750 °C with a duration of 10 min, allowed to cool and then weighed.

The sample of one gram of coal was then placed in the crucible. The crucible with the sample was placed in the muffle at a temperature of 105 °C for a time of 2 h, allowed to cool and then weighed. Then the sample was placed again with the same temperature, but now with a duration of 1 h and weighed.

Then the muffle was set at a temperature of 950 °C and the crucibles were placed for 11 min, allowed to cool and weighed.

To convert the coal into ash, the crucibles with the samples were placed in the muffle with a temperature of 750° C for a time of 6 h, it was allowed to cool and weighed, then it was placed again, but now only lasted 1 h inside the muffle and was weighed again.

The ash was placer in plastic container s and then analyzed.

The Higher heating value was determined using the next formula:

$$\mathrm{HHV} = 354.5 * \mathrm{FC} + 170.8 * \mathrm{VM}$$

where:

HHV = Higher heating value
CF = Fixed carbon
VM = Volatile material
354.3 and 170.8 = Constant data

3 Results and Discussion

3.1 Number of Trees per Hectare

The number of trees found was statistically different ($p < 0.05$) between sites, there were also differences between diametric categories ($p < 0.001$), while the interaction site: diametric categories was statistically similar. The 5 cm diametric category was statistically different from the others, which presented 510 trees/ha, while in categories 10, 15, and 20 cm, 210, 55 and 30 trees/ha were found, respectively. The total number was 805 trees/ha which is less than 1259 trees/ha reported by Donjuán et al. (2013) and even less than 9680 and 4630 trees/ha established by González-Rodríguez et al. (2013) in typical scrub area. However, to the above, if a negative exponential trend was found in the number of individuals as the diameters increase. Figure 4 shows the mean value and standard deviation by category.

Volume of Firewood

The volume of firewood was significantly different ($p < 0.01$) between type (logs and branches), between diametric categories and type interaction: diametric category. Figure 5 shows the average volume of firewood in the two types of biomasses, in each diametric and total category. If you consider the age of 20 years of *V. farnesiana* trees from the sampled sites, produce on average 0.14 m^3/ha/year, which is less than 0.211 reported by Heya (2014) in a typical area of scrub and higher than the volume produced in planting of this species quantified at 0.061 m^3/ha/year (Heya 2014).

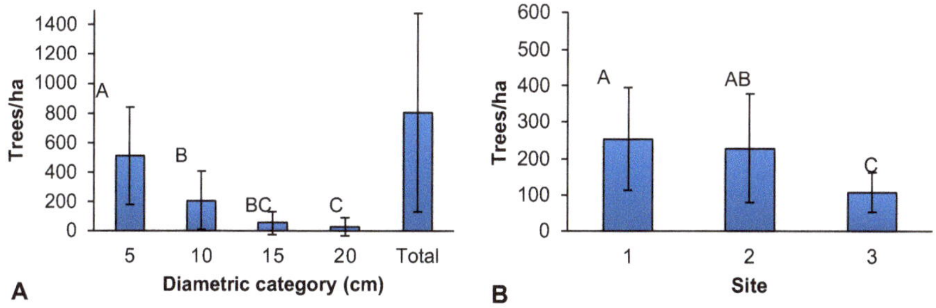

Fig. 4 (**a**) Number of trees/ha of *Vachellia farnesiana* by diametric and total category; (**b**). Number of trees/ha per site. Different letters indicate significant statistical differences ($p \leq 0.05$)

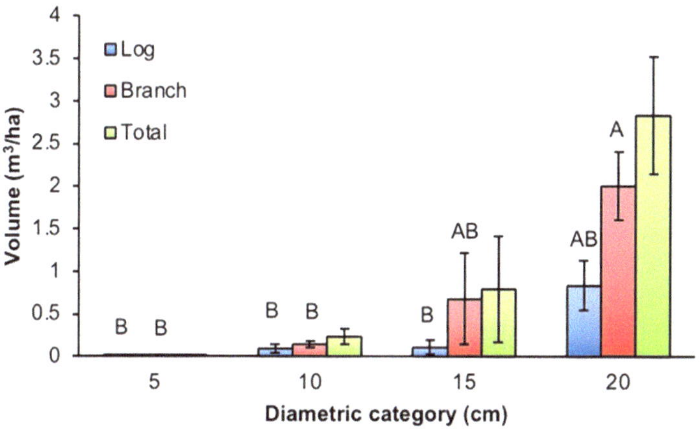

Fig. 5 Volume of logs, branches and total of *Vachellia farnesiana* distributed by diametric category. Different letters in each type of product indicate significant statistical differences ($p \leq 0.05$)

3.2 Biomass

Biomass was significantly different ($p < 0.05$) between branches and logs, and between diametric categories, as well as in the biomass type interaction: diametric category (Fig. 6). The values found are like the range of 6000–8000 Kg/ha established by Espinoza-Bretado and Návar (2005) in hilly sites and plateau.

3.3 Charcoal Yield and Characterization

The yield calculated as the charcoal mass/firewood mass ratio of *V. farnesiana* charcoal did not present statistically significant differences ($p > 0.05$) between diametric categories. The values were found within the range of 31 to 42%, which

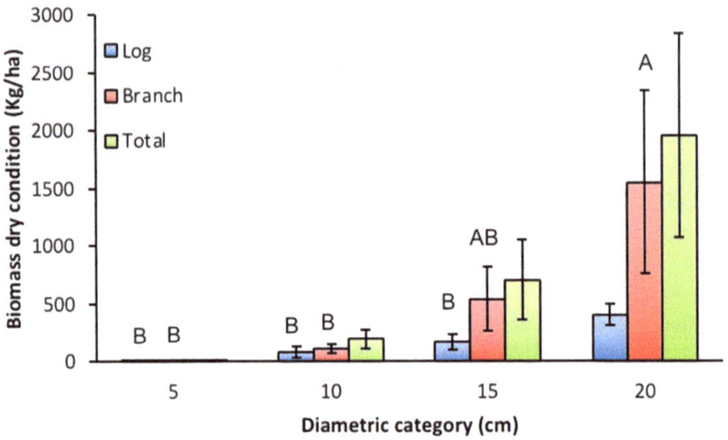

Fig. 6 Biomass of logs, branches and total of *Vachellia farnesiana* distributed by diametric category (Values in anhydrous condition). Different letters in each product type indicate significant statistical differences ($p \leq 0.05$)

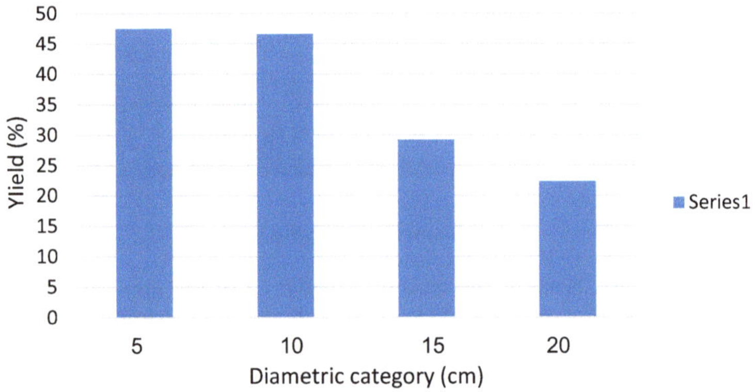

Fig. 7 Yield of *Vachellia farnesiana charcoal* when using firewood different Diametric category

is higher than the range of 26.08 to 36.69% reported by Ruiz-Aquino et al. (2019) in their work of pyrolysis under laboratory conditions with the species *Alnus acuminata*, *Arbutus xalapensis, Myrsine juergensenii, Persea longipes* and *Prunus serotina* (Fig. 7).

The characterization of coal was determined by proximal analyses. Figure 8 shows the results of moisture content, volatile material, ash, and fixed carbon

Regarding the moisture content on Fig. 8, this was statistically different ($p < 0.001$) between diametric categories. The results are in the range of 0.18 to 4.47%, which are within the specifications established by din-51749 (1989).

The volatile material was statistically different ($p < 0.05$) between diametric categories. The results are in the range of 5.56 to 8.18%, values lower than the 31.27% reported Ruiz-Aquino et al. (2019). The difference in values can be

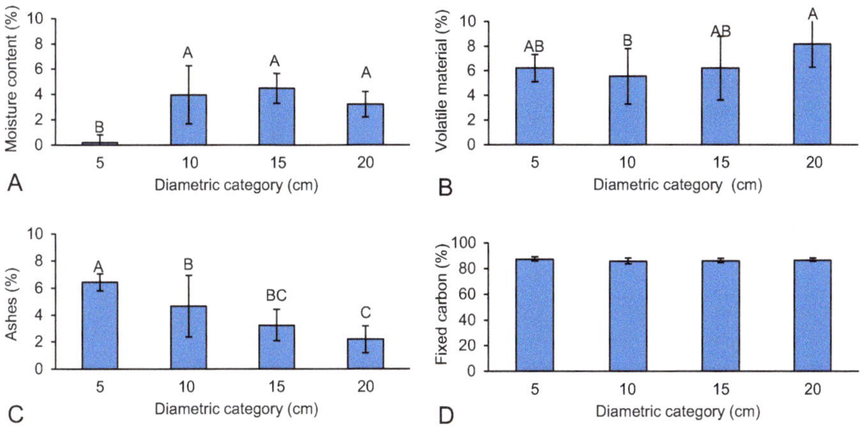

Fig. 8 Proximate analysis of charcoal from four diametric categories of *Vachellia farnesiana*. (**a**) Moisture content; (**b**) Volatile material; (**c**) Ash and (**d**) Fixed carbon. Different letters indicate significant statistical differences (p I 0.05)

attributed by using during the pyrolysis process a temperature of 750 °C which combusts much of the volatile material, considered this as a homogeneous pyrolysis process (Chin and Siddiqui 2000).

The ash content was statistically different ($p < 0.0001$) between diametric categories. The results are in the range of 2.19 to 6.43%, which is higher than the range (1.06–4.68%) reported by Ruiz-Aquino et al. (2019). The values are within the recommended by EN-1860-2 (2005) (<8%). Most diametric categories (except 5 cm) are less than 6.0% suggested by DIN-51749 (1989).

Fixed carbon was statistically similar ($p > 0.05$) between diametric categories. The results presented the range of 85.82 to 87.17%, which is higher than the minimum of 60% established by FAO (1983). The values are also higher than the 55.56–77.63% reported for charcoal made from tropical species by de Quezada et al. (2019) and the range (61.23 to 71.66%) established by Ruiz-Aquino et al. (2019).

The higher heating value was statistically similar ($p > 0.05$) between diametric categories. The range between diametric categories was from 31,038 to 31,649 KJ/kg, which is like 31,110 KJ/kg reported by Ruiz-Aquino et al. (2019) (Fig. 9).

4 Conclusions

The biomass and firewood of trees of the species *Vachellia farnesiana* present in cattle grazing areas of northeastern Mexico is generated in volumes and important characteristics to be used for energy purposes. The numbcr of trees/ha of the lower categories (5–10 cm) are considered sufficient to replace the potentially usable trees of the upper category (20 cm). The volume and quality are suitable for use as firewood and/or charcoal. The energy characteristics of this species and in the

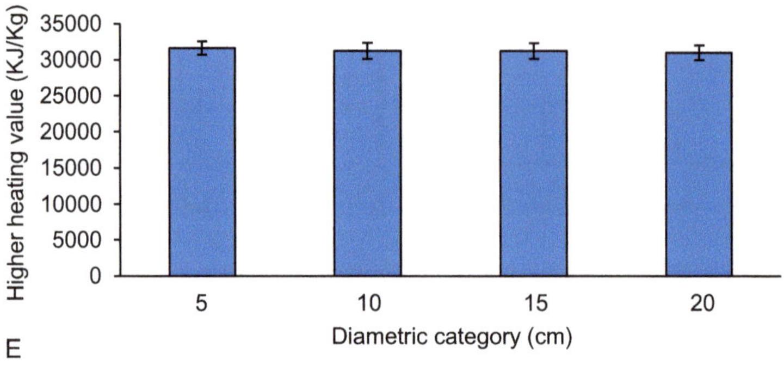

Fig. 9 Higher heating value of carbon of four diametric categories of *Vachellia farnesiana*

conditions (diameter and height) in which it has been generated are within the parameters and quality standards. Therefore, a systematic use of trees with certain characteristics of the diameter for their energy use is proposed.

References

Callejas ES, Quezada VG (2009) Los biocombustibles. El cotid., no 157, pp 75–82

Cavalcante AMB, Cox RD (2016) *Acacia farnesiana* (L.) Willd.—a potentially invasive alien species? Int J Ecol Environ Sci 42(3):209–215

Chin OC, Siddiqui KM (2000) Characteristics of some biomass briquettes prepared under modest die pressures. Biomass Bioenergy 18(3):223–228

Collard F-X, Blin J (2014) A review on pyrolysis of biomass constituents: Mechanisms and composition of the products obtained from the conversion of cellulose, hemicelluloses and lignin. Renew Sustain Energy Rev 38:594–608

DIN-51749 (1989) Testing of solid fuels- Grill charcoal and grill charcoal briquettes: Requirements and test methods. DIN, Berlin

Donjuán M, Alberto C, Alanís Rodríguez E, Jiménez Pérez J, González Tagle MA, Yamallel Y, Israel J, Cuellar Rodríguez LG (2013) Estructura, composición florística y diversidad del matorral espinoso tamaulipeco, México. Ecol Apl 12:29–34

EN-1860-2 (2005) Appliances, solid fuels, and firelighters for barbecueing - part 2: Barbecue charcoal and barbecue charcoal briquettes - requirements and test methods. The British Standards Institution European Norm, London

Espinoza-Bretado R, Návar J (2005) Producción de biomasa, diversidad y ecología de especies en un gradiente de productividad en el matorral espinoso tamaulipeco del nordeste de México. Rev Chapingo Ser Ciencias For. y del Ambient. 11, 25–31

FAO (1983) Métodos simples para fabricar carbón vegetal. Estudio FAO, Roma, Italia: Montes, 41.

González-Rodríguez H, Ramírez-Lozano RG, Cantú-Silva I, Gómez-Meza MV, Cotera-Correa M, Carrillo-Parra A, Marroquín-Castillo JJ (2013) Producción de hojarasca y retorno de nutrientes vía foliar en un matorral desértico micrófilo en el noreste de México. Rev Chapingo Ser Ciencias For del Ambient 19(2):249–262. https://doi.org/10.5154/r.rchscfa.2012.08.048

Hernández JG, Ybarra EJ (2008) Caracterización del matorral con condiciones prístinas en Linares NL, México. Ra Ximhai Rev científica Soc Cult y Desarro Sosten 4:1–22

Heya MN (2014) Bioenergy potential of shrub from native species of northeastern Mexico. Int J Agric Policy Res 2:475–483

IEA (2018) Renewables 2018, analysis and forecasts to 2023. France, Paris. https://www.iea.org/renewables2018/

Ledezma G, Wegelmy Y, Cantú Silva I, González Rodríguez H, Yáñez Díaz MI (2018) Pérdidas por intercepción de lluvia en el Matorral Espinoso Tamaulipeco bajo diferentes intensidades de raleo. Rev Mex Ciencias For 9:148–164

Molina-Guerra VM (2013) Costras biológicas del suelo en ecosistemas de pastizales del altiplano del norte de México. Universidad Autónoma de Nuevo León, Tesis de Doctorado, p 135

de Quezada JDG, Carrasco GAP, Wehenkel CA, Ángel M, Bretado E, Aquino FR, Parra AC (2019) Caracterización energética del carbón vegetal de diez especies tropicales. Revista Mexicana de Agroecosistemas 6(1):37–47

Ruiz-Aquino F, Ruiz-Ángel S, Santiago-García W, Fuente-Carrasco ME, Sotomayor- Castellanos JR, Morelia M, Carrillo-Parra MA (2019) Energy characteristics of wood and charcoal of selected tree species in Mexico. Wood Res 64:71–82

Santos RC, de Carneiro ACO, Trugilho PF, Mendes LM, Carvalho AMML (2012) Análise termogravimétrica em clones de eucalipto como subsídio para a produção de carvão vegetal. Cerne 18(1):143–151

Vogel E, Wolf F (2012) Características del carbón vegetal en algunas especies madereras del noreste de México. Rev Mex Ciencias For 11(59)

Macromycetes Associated with Three Types of Vegetation in the Municipality of Rayones, Nuevo León

Karen Elisama Rivera Luna, Fortunato Garza-Ocañas, and Inés Yañez Díaz

Abstract The municipality of Rayones is located at the south part of the state of Nuevo León, its vegetation is composed of 40% temperate forests, and they have a high diversity of species of macromycetes are associated. In this study, collection monitoring was carried out to know the diversity and distribution of macromycetes in established plots in the forest at different altitudes. Results showed 92 species belonging to 9 classes, 18 orders, 34 families and 60 genera of the phyllum Basidiomycota, Ascomycota and Mycetozoa. The families Agaricaceae and Polyporaceae had the highest number of species with 9 each, followed by the families Inocybaceae, Marasmiaceae and Xylariaceae with 6 species for each of them; at the level of genera Crepidotus (6), *Ganoderma* (5), *Leucocoprinus* (4), *Peniophora* (4) and *Trametes* (3) had a greater presence in all the sampled localities. 14 toxic and 6 medicinal species were reported. Saprotrophic, Parasitic and Mycorrhizal.

Keywords Diversity · Macromycetes · Temperate forest · Altitudinal heights

1 Introduction

The municipality of Rayones is located in the south of the state of Nuevo León and is covered in 40% of temperate forests (INEGI 2009) in which oak species are present such as e.g. *Quercus rysophylla* Weath., *Q. laeta* Liebm., *Q. polymorpha* Schltdl. & Cham., and *Q. canbyi* Trel., in addition to strawberry tree i.e. *Arbutus xalapensis* Kunth.; and some species of pines such as *Pinus teocote* Schltdl. & Cham., *P. pseudostrobus* Lindl., *P. arizonica* Engelm., *P. ayacahuite* C. Ehrenb. ex Schltdl.

K. E. R. Luna
Laboratorio de Micología, Facultad de Ciencias Forestales Universidad Autónoma de Nuevo León, Linares, Nuevo León, Mexico

F. Garza-Ocañas (✉) · I. Y. Díaz
Laboratorio de Suelos, Facultad de Ciencias Forestales, Universidad Autónoma de Nuevo León, Monterrey, Nuevo León, Mexico

F. Garza-Ocañas (ed.), *Sustainable Management of Natural Resources*, Earth and Environmental Sciences Library, https://doi.org/10.1007/978-3-031-33394-1_5

and P. hartwegii Lindl.; and in areas of greater aridity is located the stone pine i.e. *P. cembroides* Zucc (Návar 2010). In forests, biological diversity includes vegetation, fauna, fungi, and microorganisms and all of them together are integrated and synchronized in the functioning of the forest and this complexity of nutrition networks provides environmental services (ONU 2011). It is currently known that each type of vegetation in the world is characterized by a diversity of species of saprobial fungi, parasites, pathogens and mycorrhizal (Garza et al. 2002). In addition, there is a very close relationship between the diversity of plant species, their pristine condition or even poor management, with the diversity and richness of fungal species present (Garza et al. 2002). Macroscopic fungi play a very important role in the maintenance of forests, since they are degraders of organic matter and intervene in the geochemical cycles of several macro and microelements (Pardavé and Terán 1999). The work carried out in Mexico on the macromycetes of different types of oak forests are few, although they are considered one of the richest floristically (Garza et al. 1985; Zarco 1986). Macromycetes are filamentous, eukaryotic, achlorophylic, heterotrophic organisms, which reproduce asexually and sexually by means of spores, they also have a cell wall consisting mainly of chitin, they are nourished by absorption and form sporocarps whose size can vary from 1mm and in some cases as in *Termitomyces* spp., and *Macrocybe titans* their size can reach 80 cm or more and weigh more than 20 kilos, these characteristics are located as macromycetes since they are visible to the human eye (García-Jiménez 2013).

The kingdom Fungi is represented in four Phylum: Zygomycota, Glomeromycota, Ascomycota and Basidiomycota (Courtecuisse and Duhem 1995; Petersen and Læssøe 2014; García-Jiménez 2013). Macromycetes are particularly active in the surface part of the forest floor and as the depth increases their activity progressively decreases and that of micromycetes increases. For the growth development and reproduction of fungi to be carried out properly, optimal abiotic conditions such as temperature (25–35 °C), high relative humidity (70%), acidic pH (5.5–6.5) as well as nutrient availability are required (Marcano 1998; Pazos 2007).

The fungal diversity in Mexico is very large, due to its biogeographic and climatic position. In addition, its intricate orography favors a great variety of climates, which causes the complex vegetal mosaic that covers the national territory (Guzmán 1995). More than 6000 species of fungi are currently known in Mexico, this figure is distributed in approximately 2000 micromycetes and 4000 macromycetes including in the latter lichens and myxomycetes (Guzmán 1995; Buscardo et al. 2009; Pardavé et al. 2007), although it is estimated that in our country there could be more than 200,000 species of fungi and more than 3 million in the world (Hawksworth and Lückiing 2017; Guzmán 1998).

Mushroom Habit

Symbiosis was originally defined by Anton de Bary (1879) as life together between different organisms and were divided into mutualism, parasitism, and saprotrophs (Marín et al. 2015). Fungi live on organic matter, whether living or dead, which they degrade and can also obtain mineral elements from the soil for food. The species that develop on living matter are parasitic, pathogenic, mycorrhizal and saprotrophs (Guzmán et al. 1993).

Parasitic fungi are those that develop within the cells of plants or animals causing the death of these and sometimes of the whole organism according to the intensity of parasitism (Guzmán et al. 1993).

Pathogenic fungi are characterized by their rapid growth with high production of enzymes that destroy the cell wall of the parts of plants on which they grow e.g. roots, branches, fruits, seeds. Treatises on this group of fungi report that they are highly "virulent" in their activity and are of great importance both at the forestry and agricultural levels (Agrios 2005). This is the case of mutualistic symbioses, the example of mycorrhizae (from the Greek mykes: fungus and rhyza: root) formed by species of fungi and the roots of most families of terrestrial plants (Marín et al. 2015). In them the plant receives mineral nutrients and water provided by the fungus, while the latter obtains sugars product of photosynthesis (Marín et al. 2015).On the other hand, there are saprophytic fungi that grow in the soil, trunks or on agricultural waste, and these fungi degrade organic matter to achieve their growth and reproduction (Guzmán et al. 1993). Currently the information on diversity of macromycetes in the state of Nuevo León is incomplete for some municipalities, such is the case of the municipality of Rayones, Nuevo León.

The present study reports for the first time the diversity of macromycetes species associated with various types of vegetation in the municipality of Rayones. The studies reported by Marmolejo (2000) compare the richness and diversity of species in two types of vegetation (Pine Forest and Oak Forest) located in the Chipinque Ecological Park (PECh) municipality of Garza García and "Bosque Escuela" municipality of Iturbide. He recorded a total of 115 species, 86 for the school Forest and 66 for Chipinque. In that it was shown that the oak forest in the school Forest has a greater diversity of macromycetes compared to the locality of the PECh, this difference reported that perhaps it is due to the levels of air pollution with which the forest is constantly exposed at the PECh due to its proximity to the large city.

2 Methodology

The data collection was carried out in four sites located 20 km from the municipal seat by the Montemorelos-Rayones highway that connects both municipalities (Fig. 1), along the Sierra Madre Oriental with vegetation of Submontane Scrub in the lowest part and Temperate Forest in the highest.

The study sites are located between 767 m above sea level (sites 1 and 2) and 1403 m above sea level (sites 3 and 4), in them are located plant species such as e.g. *Chiococca pachyphylla* Wernham, *Croton suaveolens* Torr., *Rhus pachyrrhachis* Hemsl., *Zanthoxylum fagara* L. Sarg., *Eysenhardtia texana* Scheele, *Arbutus xalapensis* H.B.K., *Prunus serotina* Ehrh., *Quercus rysophylla* Weath., *Q. polymorpha* Cham. & Schl., *Q. laeta* Liebm., *Q. virginiana* (Small) Sarg., *Q. canbyi* Trel., *Q. laceyi* Small, *Decatropis bicolor* (Zucc.) Radlk., *Litsea*

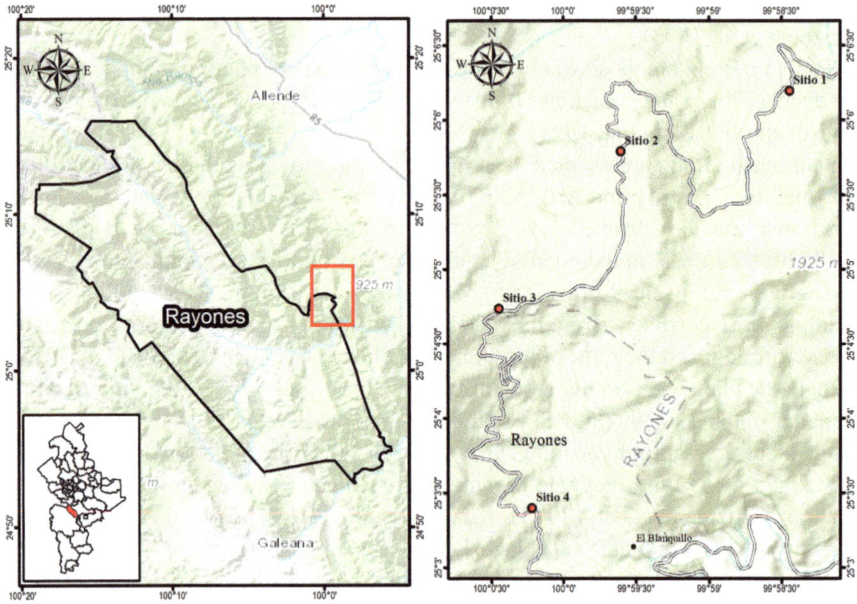

Fig. 1 Location of sampling sites

novoleontis Bartlett, *Cercis canadensis* L., *Pinus teocote* Schl. & Cham (Fig. 2). The collections of macromycete species were carried out in the rainy season and consisted of locating accessible sites due to the great slope, at different altitudes and the random sampling method was used covering areas of 1 ha of vegetation. The collection was carried out following the procedures of field mycology and the species were taken to the laboratory in coolers to proceed with the dehydration and study of the corresponding microscopy to carry out the identification of the species (Garza et al. 2002; Marmolejo 2000).

3 Results

A total of 140 specimens were collected and recorded for the municipality making a total of 92 species.

These belong to 34 families of the Phylum Ascomycota, Basidiomycota and Myxomycota. The species with the highest number of individuals collected are *Stereum ostrea* (Blume & T. Nees) Fr. and *Phellinus gilvus* (Schwein.) Pat., both species are very well distributed along the altitudinal levels studied. The registered species are presented hierarchically in the following list using the criteria established by Kirk et al. (2008).

Fig. 2 Vegetation present at each collection site. (**a**) Submontane scrub. (**b**) Oak Forest. (**c**) Oak -Pine Forest. (**d**) Oak-Pine Forest

Phylum: Ascomycota Subphylum: Pezizomycotina
Clase: Dothideomycetes
Order: Patellariales
Family: Petellariaceae
Rhytidhysteron rufulum (Spreng.) Speg.
Clase: Leotiomycetes
Order: Helotiales
Family: Pezizellaceae
Calycina citrina (Hedw.) Gray
Clase: Sordariomycetes
Order: Xylariales
Family: Hypoxylaceae
Annulohypoxylon thouarsianum (Lév.) Y.M. Ju, J.D. Rogers & H.M. Hsieh
Annulohypoxylon truncatum (Starbäck) Y.M. Ju, J.D. Rogers & H.M. Hsieh
Daldinia *concentrica*
Hypoxylon sp.
Family: Graphostromataceae
Biscogniauxia atropunctata (Schwein.) Pouzar

(continued)

Phylum: Basidiomycota Subphylum: Agaricomycotina
Clase: Dacrymycetes
Order: Dacrymycetales
Family: Dacrymycetaceae
Dacrymyces chrysospermus Berk. & M.A. Curtis Clase: Tremellomycetes
Order: Tremellales Family: Tremellaceae
Tremella lutescens Lloyd.
Clase: Agaricomycetes
Order: Agaricales
Family: Agaricaceae
Cyathus poeppigii Tulasne & C.Tulasne Especie: *Lepiota cristata* (Bolton) P. Kumm.
Leucoagaricus sp.
Leucocoprinus brebissonii (Godey) Locq.
Leucocoprinus fragilissimus (Berk. & M.A. Curtis) Pat. Especie: *Leucocoprinus* sp. 1
Leucocoprinus sp. 2
Family: Crepidotaceae
Crepidotus applanatus (Pers.) P.Kumm.
Crepidotus mollis (Shaeff.) Staude
Crepidotus sp.
Crepidotus sp. 2
Crepidotus sp. 3
Crepidotus sp. 4
Family: Cyphellaceae
Chondrostereum purpureum (Pers.) Pouzar
Family: Entolomataceae
Clitopilus sp.
Entoloma sp. 1
Entoloma sp. 2
Family: Hymenogastraceae
Gymnopilus sp.
Gymnopilus sp. 2
Family: Lycoperdaceae
Lycoperdon curtisii Berk. Especie:
Lycoperdon echinatum Pers
Family: Marasmiaceae
Campanella sp.
Gerronema strombodes (Berk. & Mont.) Singer
Gerronema sp.
Tetrapyrgos nigripes (Fr.) E. Horak Family: Mycenaceae
Hydropus sp
Mycena pura (Fries) Quélet
Mycena sp.
Panellus stipticus (Bull.) P. Karst.

(continued)

Family: Omphalotaceae
Gymnopus dryophilus (Bull. & Fr.) Murrill
Gymnopus fusipes (Bull.) Quél.
Family: Physalacriaceae
Cylindrobasidium sp.
Pluteus cervinus (Schaeff.) Kumm.
Pluteus sp.
Family: Psathyrellaceae
Coprinellus micaceus (Bull.) Fr.
Psathyrella sp.
Psathyrella sp. 2
Family: Schizophyllaceae
Schizophyllum commune Fr.
Schizophyllum umbrinum Berk.
Order Auriculariales
Family: Auriculariaceae
Exidia glandulosa (Bull.) Fr.
Order Boletales
Family: Diplocystidiaceae
Astraeus hygrometricus (Pers.) Morgan Order Cantharellales
Family: Hydnaceae
Sistotrema confluens Pers.
Order Corticiales
Family: Corticiaceae
Corticium sp.
Order Geastrales
Family: Geastraceae
Geastrum minimum Schwein.
Geastrum pectinatum Pers.
Order Hymenochaetales
Family: Hymenochaetaceae
Fuscoporia ferruginosa (Schrad.) Murrill
Phellinus rimosus (Berk.) Pilát
Phellinus gilvus (Schwein.) Pat.
Phellinus sp.
Order Polyporales
Family: Fomitopsidaceae
Neoantrodia serialis (Fr.) Audet
Daedalea xantha (Fr.) A. Roy & A.B. De
Family: Irpicaceae
Byssomerulius corium (Pers.) Parmasto
Byssomerulius incarnatus (Scwein.) Gilb
Irpex lacteus (Fr.) Fr.
Vitreoporus dichrous (Fr.) Zmitr.

(continued)

Family: Meruliaceae
Merulius tremellosus Schrad.
Phlebia sp.
Family: Phanerochaetaceae
Phlebiopsis gigantea (Fr.) Jülich
Family: Podoscyphaceae
Abortiporus biennis (Bull.) Singer
Family: Polyporaceae
Daedaleopsis confragosa (Bolton) Pers.
Ganoderma applanatum (Pers.) Pat.
Ganoderma australe (Fr.) Pat.
Ganoderma curtisii (Berk.) Murrill
Ganoderma resinaceum Boud.
Lentinus crinitus (L.) Fr. Especie:
Pseudofavolus tenuis (Fr.) G. Cunn.
Pycnoporus sanguineus (L.) Murrill
Trametes maxima (Mont.) A. David & Rajchenb.
Trametes trogii Berk.
Trametes villosa (Sw.) Kreisel
Truncospora mexicana Vlasák, Spirin & Kout
Family: Steccherinaceae
Junghuhnia nitida (Pers.) Ryvar.
Steccherinum ochraceum (Pers. ex J.F. Gmel.) Gray
Order Russulales
Family: Peniophoraceae
Peniophora cinerea (Pers.) Cooke
Peniophora albobadia (Schwein.) Boidin
Peniophora laeta (Fr.) Donk
Family: Stereaceae
Stereum ostrea (Blume & T. Nees) Fr.
Phylum: Myxomycota Order
Physarales
Family: Physaraceae
Fuligo septica (L.) F.H. Wigg.
Order Stemonitales
Family: Stemonitidaceae
*Stemonitis fusc*a Roth
Order Trichiales
Family: Trichiaceae
Hemitrichia serpula (Scop.) Rostaf.
Order Protosteliales
Family: Ceratiomyxaceae
Ceratiomyxa fruticulosa (O.F. Müll.) T. Macbr.

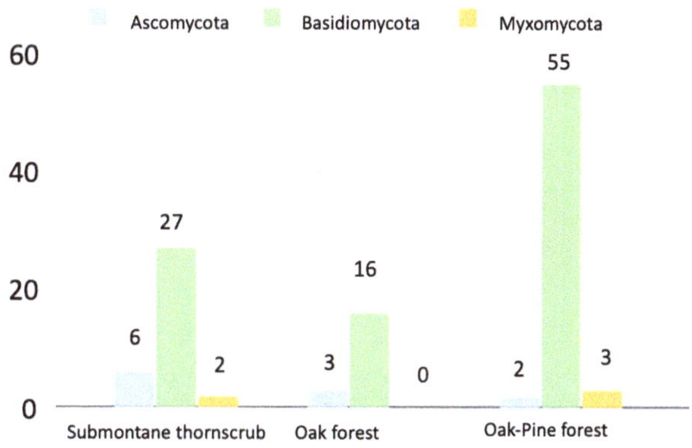

Fig. 3 Number of species of Phylum Ascomycota, Basidiomycota and Myxomycota at each vegetation type

Table 1 Taxonomic levels in the different phylum and the percentage they represent in terms of the total species collected

Phylum	Class	Order	Families	Genera	Species	%
Ascomycota	3	3	3	6	8	8.69
Basidiomycota	4	11	27	50	80	86.95
Myxomycota	2	4	4	4	4	4.34
Total	**9**	**18**	**34**	**60**	**92**	**100**

The 92 species identified correspond to 9 classes, 18 orders, 34 families and 60 genera. The Phylum Basidiomycota has the largest number of species with 80 of them and correspond to 11 orders, 27 families and 50 genera. The Phylum Ascomycota is found in a smaller proportion with 8 species. For the Phylum Myxomycota 4 species were identified (see Fig. 3).

The 92 species identified correspond to 9 classes, 18 orders, 34 families and 60 genera. The Phylum Basidiomycota has the highest number of species with 86.95% followed by the Phylum Ascomycota with 8.69% and the Phylum Myxomycota with 4.34% (see Table 1).

3.1 *Families*

In general, the families with the highest number of species were Agaricaceae and Polyporaceae with 9 each, followed by Inocybaceae (6), Marasmiaceae (6), Xylariaceae (6), and Ganodermataceae (5). The families Hymenochaetaceae, Meruliaceae and Peniophoraceae with 4 species each; and the families with the lowest number of species were: Astraeaceae, Auriculariaceae, Ceratiomyxaceae, Corticiaceae, Dacrymycetaceae, Heliotaceae, Hydnaceae, Hysteriaceae, Meripilaceae, Physalacriaceae, Physaraceae, Stemonitidaceae, Tremellaceae, Trichiaceae, Tricholomataceae with one species each (see Fig. 4).

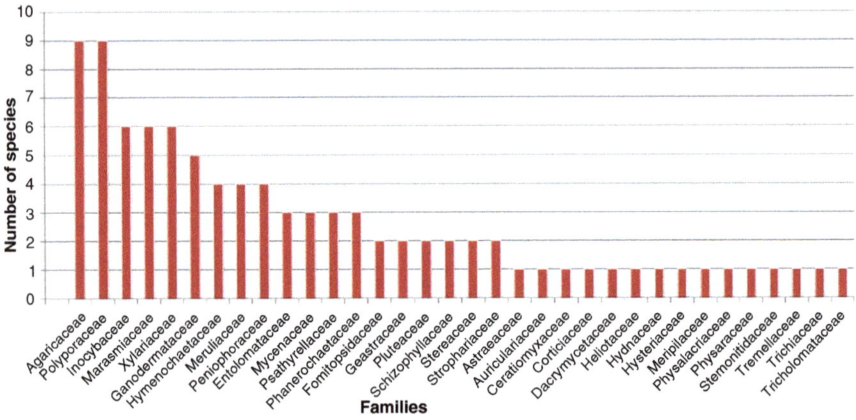

Fig. 4 Number of species from each family of macromycetes at Rayones Nuevo León

3.2 Genera

The genera with the highest number of species present in the study area were *Crepidotus (6), Ganoderma (5), Leucocoprinus (4), Peniophora (4), Trametes (3). The genera Annulohypoxylon, Antrodia, Byssomerulius, Entoloma, Fulvifomes, Geastrum, Gerronema, Gymnopilus, Hypoxylon, Lycoperdon, Mycena, Pluteus, Psathyrella, Schizophyllum,* and *Stereum* presented 2 species. The rest of the genera found: *Abortiporus, Astraeus, Biscogniauxia, Bisporella, Campanella, Ceratiomyxa, Clitopilus, Collybia, Coprinellus, Corticium, Cyathus, Cylidrobasidium, Dacrymyces, Daedalopsis, Daldinia, Exidia, Fuligo, Fuscoporia, Gloeoporus, Gymnopus, Hemitrichia, Hexagonia, Hydropus, Irpex, Junghuhnia, Lentinus, Lepiota, Leucoagaricus, Merulius, Panellus, Phellinus, Phlebia, Pycnoporus, Rhytidhysteron, Sistotrema, Steccherinum, Stemonitis, Tetrapyrgos, Tremella, and Truncospora* had only one species and do not appear in the graph (see Fig. 5).

3.3 Ecological Parameters

Most of the species (i.e. 60) were found in Oak-Pine vegetation with altitudes ranging from 1321 to 1403 m of altitude, of which 47 were collected exclusively in this type of vegetation. For the Submontane Scrub, 34 species are reported with 25 exclusives; this was followed by the Oak Forest which presented 20 species and 9 exclusive (Table 2).

3.4 Species with Wide Distribution

The species *Annulohypoxylon thouarsianum, Hypoxylon truncatum, Tremella lutescens, Phellinus gilvus, Ganoderma applanatum, Ganoderma resinaceum, Gloeoporus dichrous, Lentinus crinitus, Pycnoporus sanguineus, Trametes trogii,*

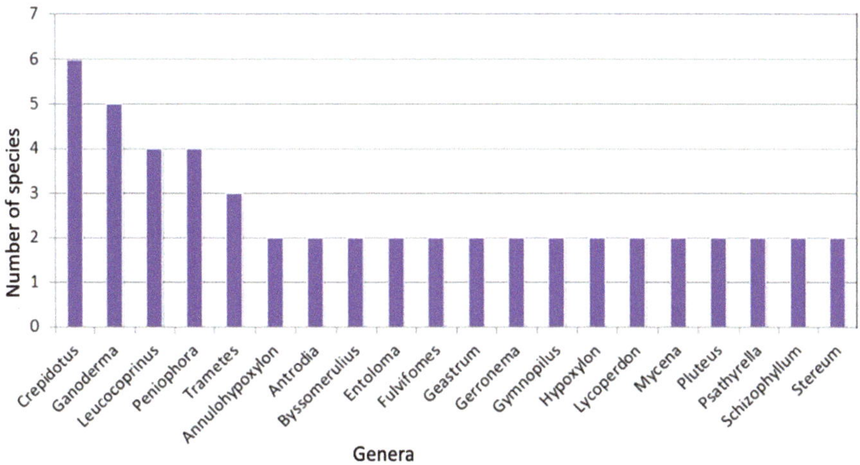

Fig. 5 Number of species per genus collected in the municipality of Rayones, Nuevo Leon

Table 2 Species collected by type of vegetation and exclusive species (U) for each one of them

Vegetation type	Altitude (m)	Total shared species	Unique species per vegetation
Submontane thornscrub	767	34	25
Oak Forests	1243	20	9
Oak-Pine Forests	1321–1403	60	47
		114	81

Stereum ostrea and *Stemonitis fusca* are widely distributed along altitudinal levels ranging from 767 m to 1403 m. The following Table 3 shows the species indicating the type of vegetation at which they were recorded.

3.5 *Edibility*

The edibility and use of the species was determined, none were identified as edible, 14 are reported in the literature as toxic and 6 with medicinal properties, example are the species of the Ganoderma (Fig. 8) since according to Ríos-Cañavate (2008) they have anticancer, immunomodulatory, anti-inflammatory, hypoglycemic, lipid-lowering and hepatoprotective properties (Table 4).

3.6 *Habit of the Species*

Of the 92 species studied for the municipality, 79 have a saprotrophic, xylophagous growth habit since they grow on wood and represent 85.86% of the total species; only 13 species are parasitic (See Table 5).

Table 3 List of macromycete species present in each type of vegetation. (I) Submontane scrub; (II) Oak Forest; (III) Oak-Pine Forest

No.	Species	Collected at different vegetation type		
		I	I	III
1	*Rhytidhysteron rufulum*	x		
2	*Calycina citrina*	x		
3	*Annulohypoxylon thouarsianum*[a]	x	x	x
4	*Annulohypoxylon truncatum*	x		
5	*Daldinia* sp.			x
6	*Hypoxylon truncatum*[a]	x	x	
7	*Hypoxylon* sp.	x		
8	*Biscogniauxia atropunctata*		x	
9	*Dacrymyces chrysospermus*	x		
10	*Tremella lutescens*[a]		x	x
11	*Cyathus poeppigii*			x
12	*Lepiota cristata*			x
13	*Leucoagaricus sp.*			x
14	*Leucocoprinus brebissonii*			x
15	*Leucocoprinus fragilissimus*			x
16	*Leucocoprinus* sp. 1			x
17	*Leucocoprinus* sp. 2			x
18	*Lycoperdon curtisii*			x
19	*Lycoperdon echinatum*			x
20	*Clitopilus* sp.			x
21	*Entoloma* sp. 1			x
22	*Entoloma* sp. 2			x
23	*Crepidotus applanatus*			x
24	*Crepidotus mollis*		x	
25	*Crepidotus* sp.	x		
26	*Crepidotus* sp. 2		x	
27	*Crepidotus* sp. 3			x
28	*Crepidotus* sp. 4			x
29	*Campanella* sp.		x	
30	*Gerronema strombodes*			x
31	*Gerronema* sp.			x
32	*Gymnopus dryophilus*			x
33	*Hydropus* sp.			x
34	*Tetrapyrgos nigripes*			x
35	*Mycena pura*			x
36	*Mycena* sp.			x
37	*Panellus stipticus*			x
38	*Cylindrobasidium* sp.	x		
39	*Pluteus cervinus*		x	
40	*Pluteus* sp.			x
41	*Coprinellus micaceus*			x

(continued)

Table 3 (continued)

No.	Species	Collected at different vegetation type		
		I	I	III
42	*Psathyrella* sp.			x
43	*Psathyrella* sp. 2			x
44	*Schizophyllum commune*			x
45	*Schizophyllum umbrinum*			x
46	*Gymnopilus* sp.			x
47	*Gymnopilus* sp. 2			x
48	*Gymnopus fusipes*			x
49	*Exidia glandulosa*			x
50	*Sistotrema confluens*	x		
51	*Corticium* sp.		x	
52	*Geastrum minimum*			x
53	*Geastrum pectinatum*			x
54	*Phellinus rimosus*	x		
55	*Phellinus sp.*	x		
56	*Fuscoporia ferruginosa*		x	
57	*Phellinus gilvus*[a]	x	x	x
58	*Neoantrodia serialis*	x		
59	*Daedalea xantha*	x		
60	*Ganoderma applanatum*[a]	x	x	x
61	*Ganoderma australe*			x
62	*Ganoderma curtisii*		x	
63	*Ganoderma resinaceum*[a]		x	x
64	*Ganoderma sessile*	x		
65	*Abortiporus biennis*			x
66	*Irpex lacteus*	x		
67	*Vitreoporus dichrous*[a]	x	x	
68	*Phlebia tremellosa*			x
69	*Phlebia* sp.			x
70	*Byssomerulius corium*	x		
71	*Byssomerulius incarnatus*	x		
72	*Steccherinum ochraceum*			x
73	*Daedaleopsis confragosa*			x
74	*Pseudofavolus tenuis*	x		
75	*Junghuhnia nitida*	x		
76	*Lentinus crinitus*[a]	x	x	
77	*Pycnoporus sanguineus*[a]	x		x
78	*Trametes maxima*	x		
79	*Trametes trogii*[a]	x	x	
80	*Trametes villosa*	x		
81	*Truncospora mexicana*	x		
82	*Peniophora cinerea*			x

(continued)

Table 3 (continued)

No.	Species	Collected at different vegetation type		
		I	I	III
83	*Peniophora albobadia*	x		
84	*Phlebiopsis gigantea*	x		
85	*Peniophora laeta*	x		
86	*Stereum ostrea*[a]	x	x	x
87	*Chondrostereum purpureum*		x	
88	*Astraeus hygrometricus*			x
89	*Fuligo septica*	x		
90	*Stemonitis fusca*[a]	x		x
91	*Hemitrichia serpula*			x
92	*Ceratiomyxa fruticulosa*			x

[a]Species that are widely distributed along the altitudinal levels

Table 4 Toxic and medicinal species registered at the locations

No	Toxic species	Medicinal species
1	*Leucoagaricus* sp.	*Ganoderma applanatum*
2	*Leucocoprinus brebissonii*	*Ganoderma australe*
3	*Leucocoprinus fragilissimus*	*Ganoderma curtisii*
4	*Leucocoprinus* sp. 1	*Ganoderma resinaceum*
5	*Leucocoprinus* sp. 2	*Ganoderma sessile*
6	*Clitopilus* sp.	*Pycnoporus sanguineus*
7	*Entoloma* sp. 1	
8	*Entoloma* sp. 2	
9	*Gymnopus dryophilus*	
10	*Mycena pura*	
11	*Mycena* sp.	
12	*Gymnopilus* sp. 1	
13	*Gymnopilus* sp. 2	
14	*Gymnopus fusipes*	

Table 5 Habit of the macromycetes registered and the percentage they represent at the municipality of Rayones, Nuevo León

Habit	No. of species	%
Saprotrohic	79	85.86
Parasite	13	14.13

3.7 *Parasitic and Hyper Parasitic Species*

The species recorded as parasites are *Bisporella citrina, Campanella sp., Daedalopsis confragosa, Fulvifomes rimosus, Fulvifomes sp., Fuscoporia ferruginosa, Ganoderma applanatum, Ganoderma australe, Ganoderma curtisii, Ganoderma resinaceum, Ganoderma sessile, Junghuhnia nitida*, and *Phellinus gilvus* ethese species were found growing on species of the genus Quercus. In the

Fig. 6 Hyperparasitic species recorded on sporocarps of *Phellinus* gilvus. (**a**) *Bisporella citrine*. (**b**) *Campanella* sp

case of *Bisporella citrina* and *Campanella* sp., they are hyperparasite species since they were collected growing on sporocarps of *Phellinus gilvus* (Fig. 6).

4 Discussion

For the municipality of Rayones, Nuevo León, 92 species of macromycetes are reported for the first time, most of the species were found in saprotrophic habit (79), growing mainly on wood and substrates composed of woody tissues. Only 13 species presented parasitic habit, in addition *Bisporella citrina* and *Campanella* sp. are recorded as hyperparasites (Boosalis 1964; Mueller et al. 2004) as they were found growing on sporocarps of the species *Phellinus gilvus*.

The diversity of macromycete species is influenced by the type of vegetation, altitude, and humidity of the area. Thus, the forests of Encino-Pino are located from 1200 to 1400 m of altitude and presented a greater number of species (60), as well as exclusive species for this type of vegetation with 46, compared to the Submontane Scrub with 34 species and 25 exclusive and is located at heights of 700 m. In addition, the Oak-Pine Forest has the necessary substrate and conditions, such as humidity ranging between 43 and 55% per year (Nixon 1993), all this for the development of most species of macromycetes, both saprotrophic and parasitic registered. In this study no species of mycorrhizal fungi were recorded, this can be explained because when the collections were carried out, the rains that occurred were not the most important of the season and in successive explorations the humidity disappeared quickly due to the high temperatures.

Knowledge about macromycete diversity for the state of Nuevo León is very poor. More work needs to be done on quadrants located throughout the state and work on the inventory of macromycete species. In order to develop knowledge on macromycetes species and their distribution, the studies carried out so far show that temperate forests have the greatest species richness. The results obtained contribute to the knowledge of the species de macromycetes of the municipality of Rayones, Nuevo León.

The number of species reported can increase considerably with more intensive sampling and in years with more rainfall.

Appendix: Some of the Species Collected and Registered at Rayones, Nuevo León

Fig. 7 Some species of macromycetes representative of the study areas

Fig. 8 Some species of macromycetes representative of the study areas

Fig. 9 Some species of macromycetes representative of the study areas

References

Agrios GM (2005) Fitopatología. Editorial Limusa, Segunda Edición. 821pp

Boosalis MG (1964) Hyperparasitism. Annu Rev Phytopathol 2:363–376

Buscardo E, Rodríguez S, De Angelis P, Freitas H (2009) Comunidades de hongos ectomicorrícicos en ambientes propensos al fuego: Compañero esenciales para el restablecimiento de pinares mediterráneos. Ecosistemas 18(12):55–63

Courtecuisse R, Duhem B (1995) Mushrooms & toadstools of Britain and Europe. Harper Collins Publishers. 479pp.

De Bary AH (1879) Die erscheinung der symbiose. Vortrag, gehalten auf der versammlung deutscher naturforscher und aerzte zu cassel. Verlag von Karl J. Trübner, Leipzig-Berlin

García-Jiménez J (2013) Diversidad de macromicetos en el estado de Tamaulipas, México. Tesis de Doctorado. Facultad de Ciencias Forestales. Universidad Autónoma de Nuevo León, Linares Nuevo León, México. 254pp

Garza F, García J, Castillo J (1985) Macromicetos asociados al bosque de *Quercus rysophylla* en algunas localidades del centro del estado de Nuevo León. Rev Mex De Micol 2:197–205

Garza F, García J, E. Estrada y H. Villalón. (2002) Macromicetos, ectomicorrizas y cultivos de *Pinus culminicula* en Nuevo Léon. Ciencia UANL 5:2

Guzmán G (1995) La diversidad de hongos en México. Ciencia. 39:52–57

Guzmán G (1998) Inventorying the fungi of Mexico. Biodiversity and Conservation. 7:369–384

Guzmán G, Mata G, Salmones D, Soto C, Guzmán L (1993) El cultivo de los hongos comestibles con especial atención a especies tropicales y subtropicales en esquilmos y residuos agroindustriales. Primera Edición. Instituto Politécnico Nacional. 245pp

Hawksworth DL, Lückiing R (2017) Fungal diversity revisited 2.2 to 3.8 million species. Microbiol Spectr 5(4):FUNK-0052-2016. https://doi.org/10.1128/microbiolspec.FUNK-0052-2016

INEGI (2009) Enciclopedia de los municipios y delegaciones de México. En línea: http://siglo.inafed.gob.mx/enciclopedia/EMM19nuevoleon/municipios/19043a.html

Kirk PM, Cannon PF, Minter DW, Stalpers JA (eds) (2008) Ainsworth & Bisby's dictionary of the fungi. CAB International, Wallingford, p 771

Marcano V (1998) Caracterización de los microrrefugios de la Gran Sabana, Estado Bolívar, a partir del estudio ecofísico de sus comunidades de plantas inferiores y hongos. Rev Ecol Lat Am 51(1-2):21–48

Marín MA, Silva V, Linares G, Castagnino AM, Ticante JA (2015) La biodiversidad de los hongos ectomicorricicos y su importancia para la conservación del bosque en la zona poblana del Parque Nacional Malintzi. Estudios en Biodiversidad 1:180–195

Marmolejo J (2000) Diversidad fúngica en dos ecosistemas forestales del estado de Nuevo León, México. Reporte Científico No. 36. Universidad Autónoma de Nuevo León. Facultad de Ciencias Forestales. 43pp

Mueller GM, Bills GF, Foster MS (2004) Biodiversity of fungi: inventory and monitoring methods. Elsevier Academic Press, Londres, Reino Unido, p 777

Návar J (2010) Los bosques templados del estado de Nuevo León: el Manejo sustentable de los bienes y servicios ambientales. Madera Bosques 16(1):51–69

Nixon KC (1993) The genus *Quercus* in Mexico. In: Ramamoorthy TP, Bye R, Lot A, Fa J (eds) Biological diversity of Mexico: origins and distribution. Oxford University Press, New York

ONU (2011) On line: http://www.onu.org/es/events/biodiversityday/forests.shtml

Pardavé L, Flores L, Franco V, Robledo M (2007) Contribución al conocimiento de los hongos (macromicetos) de la Sierra Fría, Aguascalientes. Investigación y Ciencia. Universidad Autónoma de Aguascalientes. 37:4–12

Pardavé M, Terán M (1999) Estudio comparativo de dos comunidades de Macromicetos en el Área Protegida de Sierra Fría. Rev Investig Cienc 20:02–10

Pazos A (2007) Los hongos del ecosistema. Agrupación Micológica A Zarrota 2(6):1–18
Petersen JH, Læssøe (2014) The Norwegian Biodiversity Information Centre (NBIC) (2018). Artsnavnebasen. Version 1.14. https://doi.org/10.15468/4dd3tf
Ríos-Cañavate JL (2008) *Ganoderma lucidum*, un hongo con propiedades inmunoestimulantes. Facultad de Farmacia, Universidad de Valencia, España, Departamento de Farmacología, p 8
Zarco J (1986) Estudio de la distribución ecológica de los hongos (principales macromicetos) en el Valle de México, basado en los especímenes depositados en el Herbario. Rev. Mex. De Micol. 2:41–72

Social Capital in the State of Nuevo León as a Tool for Sustainable Forest Development

Horacio Villalón-Mendoza, Artemio Carrillo-Parra, and Angélica Judith Ocampo-Ramos

Abstract Although the state of Nuevo León is not a purely producer of forest raw materials, it is a State that industrializes them. In the State there is a wide variability of groups, there are from official to non-governmental organizations, it is therefore important to emphasize that for a long time the peasants who inhabit forest areas have organized to seek a common good. In the country's forest areas, there is an organizational culture of the population, which probably comes from pre-Hispanic times. The main objective of this research work is to determine the social capital in the state of Nuevo León and determine if it has served as a tool for the sustainable development of forest resources. It was concluded that social capital in the State forestry sector presents a positive trend in terms of the role it plays in the State's sustainable forestry development, so it is very important to monitor its behavior in the future and to be able to evaluate the rate of development that occurs and so to have the ability for to make the most pertinent decisions regarding the sector and its socioeconomic and ecological sustainability.

Keywords Education · Society · Forestry · Development

1 Introduction

The state of Nuevo León, although it is not a State that is clearly a producer of forest raw materials, is a state that industrializes them, according to the forest inventory, the forestry vocation in the state of Nuevo León, indicates that the area with timber production is of the order of 11,942 km^2, corresponding to 9% to the wooded areas in temperate areas and 9.5% to the mosques (INEGI, 2019). On the other hand and

H. Villalón-Mendoza (✉) · A. J. Ocampo-Ramos
Faculty of Forestry Sciences, Autonomous University of Nuevo León, Linares, Nuevo León, Mexico

A. Carrillo-Parra
University Juárez del Estado de Durango, Instituto de Silvicultura e Industria de la Madera, Durango, Durango, Mexico

F. Garza-Ocañas (ed.), *Sustainable Management of Natural Resources*, Earth and Environmental Sciences Library, https://doi.org/10.1007/978-3-031-33394-1_6

Treviño (2009), report that the areas of the State dedicated to forestry activities correspond to 6%. In the state there is a wide variability of groups, there are from the official ones to the CNC (National Peasant Commission), Union of Forest Ejidos, Union of milk producers, various non-governmental organizations, etc., it is therefore important to emphasize, that for a long time the peasants who inhabit the forest areas have organized themselves to seek a common good. Historically speaking, the Spanish and Creole colonizers in the last third of the sixteenth century introduced agricultural and livestock practices to the region. By 1695 the number of heads of small cattle amounted to more than half a million, and despite marking that the lands were fertile with green pastures, there was transhumance to maintain the herds (Treviño, 2009).

The Social Observatory of Mesoamerican Agriculture (OSAM) (2009) analyzes the situation of Mexican agriculture where forest resources and the same agricultural activity are involved in Mexico and mentions that: "Mexican agriculture does not end in its crop fields. In addition to the agricultural potential of Mexico (Seminis, 2018), the country has important economic options in forest areas and wildlife, because due to the presence of extensive arid areas and mountains, only 21.8 of lors 196 million hectares are planted, while, in the rest of the country, productive activities are based on the management of grasslands, forests, bodies of water and other wildlife areas. For the above, it can be said that forest areas predominate in the Mexican landscape since 56 million hectares are wooded and that they provide the peoples with food, medicines, construction materials and energy, that is why the forests are considered as the pharmacy, butcher and hardware store of many communities It is also important to emphasize that in Mexico the rains are irregular and scarce in the center and north (Conagua, 2018), with forest massifs being water catchment areas that facilitate infiltration and feed the springs and aquifers on which agriculture, industry and urban areas depend. Four fifths of the forest lands of Mexico are socially owned mainly these lands are in the hands of ejidos and communities, being the decisive forest resources for the social, economic, cultural, and political life of the 9 thousand communities and 12 million people who inhabit them and that are almost half of our rural population that are settled in the areas with forests, subtropical vegetation and in arid areas. Forest resources also occupy an important place in the protection of Mexico's great biological diversity, as well as in the permanence of soils and water sources and in the regulation of the local climate and the planet (Semarnat, 2012). Continuing with the observations of the OSAM (2009), it mentions that forest use is not depleted in commercial timber products. Our relative efficiency in the use of energy and natural resources is linked to the fact that in Mexico there are still forms of local development that give life to regional economies. Coffee growers, producers of pine or mahogany wood, chicleros, ixtle carvers, shepherds, collectors of pepper, oregano, candelilla, wax, camedora palm, among others, maintain productive activities that, in addition to being the basis of family and collective life, give the regions of the country distinctive characteristics that make them unique. The inexhaustible biological and landscape diversity that characterizes us is not a simple natural fact depends on whether local economies based on the traditional use of biological resources subsist (Martínez et al., 2015).

Unlike most OECD countries, which take advantage of between 40 and 85% of their forest production, Mexico under employs its forest resources, with less than a fifth of what can be harvested reaching formal markets. Until now, forest and wildlife policies have not given importance to developing the potential of forest and wild areas.

The OSAM (2009) emphasizes that the lack of investment in management systems results in very weak production. According to G-Bosques: "The forest production targets proposed in the National Forest Programme (NFP) were 12 million cubic meters by 2006; however, in fact, these goals have not only not been achieved, but there has been a drastic drop in national timber forest production, of more than 22%. That is why there are records where production has gone from nine million cubic meters in 1998 to less than seven million in 2005. "However, the enormous importance of forest and wildlife areas in the country, government and other sources aimed at their sustainable use are minimal, and if there are any, they are misguided. Over the past 25 years, forest policy has been erratic and persistently marginal: multiple changes in legislation, low budgets, induction of clearing and opportunity costs distorted by agricultural development policies out of balance with forest development, among other aspects, as well as the deficiency of industrial integration for post-concession conditions."

On the other hand, the weakness of the policy on forests and wilderness areas is manifested in that: "Despite having large areas of timber forest in its territory, Mexico increasingly depends on imported timber products. Over the past seven years, the forest trade balance deficit has grown steadily and sharply: from $967 million in 1997 to a dramatic four billion in 2004. This is equivalent to 20 times the federal government's forestry budget for the sector." Another measure of the untapped potential is that the average growth rate of national stocks is only one cubic meter per hectare per year, while there are consistent data in which that figure reaches 20 m, in the case of forests located on the neo-volcanic axis."

Despite the above, there is still in the forest areas of the country a culture of organizational population, which probably comes from pre-Hispanic times, where there was distribution of activities and they organized to defend their lands and fight with rival tribes. The main objective of this research work is to determine the social capital in the state of Nuevo León and determine if it has served as a tool for the sustainable development of forest resources.

2 Background

Aguirre (2015), mention that the mobilization of social capital within the poorest sectors must be implemented together with a dynamic and inclusive economic system, in addition to a broad socio-political system that is consistent with the objectives of inclusion. This strategy should be oriented to the design of formal institutions of partnership and participation. The training of beneficiaries for their own management will never achieve its objectives, unless informal socio-cultural institutions of trust, cooperation, leadership, and prestige are recognized as public policy issues. Ocampo also suggests another important issue as a cause of the failure of traditional poverty reduction programs, which include technocratic and paternalistic relationships between development agencies and the population they serve. Communal social capital complements public services in several ways. First, participation at an associative community level can be key to articulating public services with households, which is very important in programs aimed at reducing poverty.

On the other hand, the mobilization of communal social capital can contribute to making these programs more efficient and promote urban microenterprises and peasant production. This has been observed in the forestry sector of the state of Nuevo León, mainly by the responses referred to the impacts of the programs on the sector and the situation of the work environment fostered by the same programs in the forestry sector of the State.

Fukuyama (2003) and Urteaga (2013), mention that the connection between social capital and development, whose failure can be attributed to the absence of social capital. Points out that economic development requires a competent, strong and effective state; of a set of institutions that allow the change of policies, and of the adequate predisposition of social and politicians.

For a better understanding of how social capital can be used in poverty reduction, Robison, et al. (2003) mention the social capital paradigm. It includes social capital, networks, socio-emotional goods, rooted values, institutions, and power. It is because of the above that social capital can be understood as the sympathy of one person or group towards others. Also, the distribution of social capital can be described by networks. Social capital has value because of its ability to produce economic benefits and, if ignored, results in economic disadvantages. In addition, social capital can create values because it produces socio-emotional goods, which satisfy needs of the same nature. Sometimes these goods are rooted in objects. When this happens, it changes the meaning and value of these, creating values of rootedness. Institutions are the rules that order and give meaning to exchanges. When institutions acquire values of rootedness, they are more likely to be fulfilled than when their fulfillment depends on the power derived from incentives or threats.

The components of the social capital paradigm are interdependent and necessary to understand and prescribe solutions for poverty reduction.

Uphoff (2003) mentions, in his analysis of the concept of social capital, that social capital is a reserve, which produces a flow of profits and a mutually beneficial collective action. The word "social" is derived from the Latin word "friend." It identifies two kinds of social capital: structural social capital, which comes from social structures and organizations; and cognitive social capital, which consists of psychological or emotional states (García, 2011). Structural social capital facilitates mutually beneficial collective action, while cognitive capital predisposes people to such action. It also looks at how social capital cannot be measured or managed. He argues that measures of social capital have been developed that can have both predictive and explanatory value.

Structural social capital facilitates mutually beneficial collective action, while cognitive capital predisposes people to such action.

Durston (2003) defines social capital as the content of certain social relations better expressed in acts of trust, reciprocity, and cooperation, which provide greater benefits to those who have social capital than to those who do not. This author examines the importance of concepts such as kinship, friendship, and prestige, as instruments of social change. It also notes that share capital is activated only in times of need, risk, or opportunity. Finally, this author recognizes the dynamic nature of social capital when it is related to projects that require the mobilization of different allies. It also relates social capital to the problem of the unequal distribution of resources. It recognizes the influence of social capital within impoverished social groups and dominant private groups, but also in government units that contribute to "patronage" and "lobbying groups." Likewise, the author argues that policymakers will never provide effective policies for poverty reduction if they do not include social capital as a part of their internal and policy debates.

The state/civil society relationship must be re-evaluated, to recognize both positive and negative aspects of social capital labor, which facilitates or prevents collective participation and the empowerment of the poor.

Flores and Rello (2003) argue that social capital should not be confused with what it can achieve. For example, social capital is not simply a collection of norms, institutions, and networks. They argue that by confusing this difference a very common mistake is being made, which leads researchers to consider improvements in the theory and practice of social capital not because of social capital, but of other processes, thus creating false expectations in relation to this concept.

Diaz-Albertini (2003) describes how social capital can lead to sustainable development, only if vertical ties in each national society provide the poor with access to social resources. In addition, it notes that by nature the share capital is exclusive. Trust, norms and networks tend to belong to groups with very well-defined boundaries. The challenge for the use of social capital for development is the extension and lengthening of social capital networks to include the people most in need.

Sunkel (2003) mentions that social capital has to do with participation in networks, reciprocity, trust, social norms, and proactivity. It assumes that these elements are dimensions that are used in empirical research in select communities and promote opportunities for the accumulation of social capital.

On the other hand, Bebbington (2003) deals with a reflection of the various trends of environmental and socioeconomic changes in the Andes region, and a reflection of the roles of the different actors of civil society. The focus of this author refers to peasant organizations in the processes of intensification of agricultural production, and the use of resources and life strategies. It considers that social capital, which is based on social networks and associative forms of civil society, can have considerable effects on market and government functions. It also shows the impact of these networks on civil society, as well as on the rate of distribution of the social benefits of economic growth. It suggests two future options for reducing high rates of poverty and degradation. The first predicts a continuation of rural migration to urban centers, which has a direct impact on the process of decline of traditional economic organizations, structures, and formations in a region of South America. He states that this process will have a direct effect on ecological recovery and the decrease in demographic pressure. The availability of financial resources, from migrations, will result in less intensive land use. The other option relates to a significant intensification in the efficient use of natural resources, which strengthens the productivity of land and labor.

Flora and Flora (2003) and Lutz (2011) define bonded social capital as the connections between homogeneous individuals and groups, which can be familiar to each other in multiple contexts. Bridge social capital, as the authors define it, connects diverse groups within a community with groups outside the community. The authors illustrate the importance of each type of social capital and examine its relevance to development. Among these associative resources, which are important for measuring the social capital of a group or community, he cites relationships of: trust, reciprocity, and cooperation. Combining these two dimensions, it shows four different forms of social capital, ranging from restrictive social capital to expanded social capital.

3 Objectives

Among the objectives to be achieved is to carry out the analysis of the social capital present in the state of Nuevo León, trying to understand the role it plays in the promotion of sustainable development in the productive forestry sector in the entity. To determine the effects obtained with the adoption of social capital in the forest area of the state of Nuevo León. Likewise, to know the effects of social capital on the degree of conservation of forest resources.

4 Methodology

This research was carried out in the state of Nuevo León, which is in the Northeast of Mexico, located between 27° 48′ and 23° 09′ north latitude and between 98° 26′ and 101° 13′ west longitude. It has an area of 64 555 km^2 (3.3% of the national territory).

Prior to the design and implementation of the research, the main studies carried out so far in relation to social capital in Mexico and in different regions of the world whose social, economic, and cultural characteristics are like the conditions of Mexico, and which were related to social capital as a tool to produce sustainable development in communities were reviewed. With the conjunction of the information obtained, a contextual theoretical framework was built that served as a theoretical foundation at the same time as a starting point to start the present study and served to a large extent in the conclusions and that are shown at the end of the study.

As a first step of the research, it was to establish the universe of study, which was determined by the researchers according to the objectives sought, in this way, it was determined that this universe included the daily forestry activities that take place in the state of Nuevo León, which in one way or another have a contribution to the forest development of the State. It is important to mention that, in the universe of study in this research, it was the forest area and that it includes in an integral way the forest with all its components, and therefore the forests of all the types existing in the State, the timber production, as well as the wild fauna that is found in the forest ecosystems are included and that, in addition, they represent an important element for the forestry activities of the state because they are a non-timber forest resource. Non-timber elements were not classified and were considered as part of the objective of the study for the present research as a single resource. This universe was the basis of the study, the information that did not undergo any kind of subsequent modification. For the selection of the pollsters, we worked considering the characteristics of the State and in the forestry field, together with the climate of electoral proselytism that the state was at the time of conducting the surveys, the questionnaires were also applied considering the above, the selected pollsters were students with a completed degree or in the process of completion. They were prepared through a training course-workshop and information on the study and precise indications, such as: appropriate clothing and footwear; personal and institutional identification; request without pressing permission to apply the survey; not to offer any kind of solution to your problems or individual compensation; they were also asked to add at the end the most important observations and sign each questionnaire. Likewise, the fieldwork at each site was directly supervised and coordinated by at least one of the researchers responsible for the project.

Regarding the tools for analysis, two different ones were combined that can be used for the collection of information in the field. Points out that it is advisable to use two or more tools to be able to cross fundamental information, because it helps to verify the veracity of the information obtained. The quantitative and qualitative methods used to conduct the present research were basically questionnaires or surveys, individual interviews with key informants, group interviews or focus groups, previous case studies, direct observation in the field by researchers, all of which provided a comprehensive understanding of the phenomenon.For the sampling design, the probabilistic sample design was handled, which was of the proportional stratified type, which involved the prior classification of all the elements that make up the population of informants in strata (producers and technical managers). Subsequently, the selection of the sites to be sampled randomly was made. These stratified samples resulted from the collection of partial samples from each of the strata into which the study population was divided. This led to homogeneous strata. It was decided to use this design in the present research because of the advantage it has over other types of sample design, such as simple randomization, due to the complexity of the study. In addition, because greater precision is achieved in the estimates, a situation that is presented by the decrease in size in the standard error.

In the application of the probabilistic sample design of the proportional stratified type, once the number of questionnaires that were required to be applied was determined, they were distributed in proportion to the size of the strata carried out, which were selected by the procedure indicated to take a simple random sample. The above to obtain high quality information with validity and statistical confidence, since the information can be of more value, being able to be better classified according to the classifications made to make the strata. For the determination of the sample size was determined this, a questionnaire or short survey with questions, no more than 10 and several open questions, we chose to use the following statistical formula for complex studies (Rojas 1998).

Where:

n= total sample size N= probability size

Z= Confidence level: 95% (1.96 in areas under the normal curve)

E= Level of accuracy

pq= Observed variability, where "p" represents the affirmative or "adequate" and "q" represents the negative or "inadequate" responses.

To facilitate the management and effectiveness of the information obtained, the stratification of the sample population was followed. After the pre-analysis carried out to obtain the number of surveys to be carried out, a value of "n" equal to 40 surveys was obtained, distributed in a weighted and random way among the levels of the universe ("N"). Meaning a sampling intensity of 30% with an accuracy of 0.1 and a confidence level of 95%. For the formulation of the survey, two types of surveys were designed; one aimed at active forest producers and one at key informants. These were elaborated based on the proposed objectives, which were expanded with some concepts considered of importance, to improve the understanding and characterization of the situations encountered. Both surveys contained both closed and open questions and a section so that the interviewer, in addition to recording the identification and address of the interviewee (if he allowed it in forest producers), could also record the relevant and pertinent observations detected at the time of applying the survey. Both surveys required about 10 to 15 minutes each, although depending on the environment and level of trust generated between the interviewee and the interviewer it could take up to 20 minutes per survey.

The number of valid surveys obtained during fieldwork were 10 from key informants and 30 from forest producers. In the survey, the random and opportunity method was followed, which guaranteed the significance of the sample.

The individual interview between forest producers and key informants was conducted with "unstructured" interview formats. Focus groups were also established for producer groups, which were guided by a "structured" interview.

For the processing of the information, version 16.0 of the Windows SPSS was used, software widely recommended for socio-demographic analysis. The information collected with the surveys was captured in masks made for this research by specialists in this system.

For the analysis, tabulations were constructed; with its graphs included (bars and cake), from the perspectives of cases and percentage. From these, those that reliably reflected the reality studied were selected.

5 Results

5.1 *Context*

The state of Nuevo León is a state with just over 4 million inhabitants, being 2,108,609 women and 2,090,673 men, occupying the eighth place nationally, its population is distributed mostly in children up to 14 years (14%) and young adults from 15 to 29 years (13%). There are no exact data on the rural population found in the state, but it is estimated that 20% of the state's population lives in these areas.

After brief conversations with groups of producers from rural areas, fieldwork was carried out and developed with wide freedom. A minimum of 10 adults and 20 minors were contacted in each Ejido, managing to raise 10 surveys of adults (± 3 women and 7 men). The results found in this ejido, as well as the ejidos that were subsequently visited, have validity with respect to similar studies.

5.2 *The family*

The presence of family nuclei in the ejidos has full correspondence with the general result. Also, the tendency of people in rural areas is to organize, being that this organization to work revolves around the nuclear and extended family. Here the nuclear family is the hegemonic among the population dedicated to forestry activities. Of the total of the respondents, 88 percent reported living in this type of family, only 5 percent in families headed by the mother and the remaining 7% with extended family (grandparents, uncles, mainly).

5.3 *Housing*

As for home ownership, all respondents indicated that it was their own. Regarding the material of the house, despite being in forest areas, it was found that there is a predominance of those built with high-durability material. (Brick and block) with 76%, an aspect of the house that manifests itself with less gravity is that of basic services. 100% have water, 95% electricity and 30% have drainage.

Although these services solve some basic needs for quality of life, in most households there is a lack of necessary equipment, such as refrigerator, stove and washing machine, in percentages greater than 60%. However, although some of these housing characteristics reduce the condition of the quality of life of the inhabitants of rural areas, this exceeds the trend of the national sample.

5.4 Schooling and Perspectives

In the field of education, it was found that the perception of the inhabitants of rural areas with forest resources about education is highly positive. It was found that 85% of respondents can read and write.

An important element is the perception that the inhabitants have of their own future. In the state, more than half of the respondents and of age to go to study mentioned having professional aspirations, the most important being that of teacher, doctor and engineer. Less than half, he said he didn't know. This tendency of lack of aspirations to have a professional career is strengthened by the perception that there is a high pessimism about the future.

5.5 Work in Forest Areas

The work environment in which the inhabitants of forest areas in the State work can be classified as dangerous and, in many instances, it is an arduous and strenuous work. Depending on the activities they carry out within the complicated chain of work of forest use, these are usually carried out with the minimum of protective equipment. For example: if you are a wood cutter, this activity is carried out without a safety helmet, without protective boots, as well as there is a lack of protective equipment for the eyes and ears. This activity is carried out from 6 to 8 hours a day, being advised by the World Health Organization a maximum of 2 hours and breaks of 4 hours. On the other hand, the day laborer must travel distances of up to 10 km to reach the short areas and from there start their activities. If the cutters decide to stay at the cutting site for six days a week, set up a camp with tarpaulins to protect themselves from the rains, prepare their own food, in these camps there is a total lack of sanitary services, as well as first aid kits to attend emergencies. Similar conditions occur in other activities related to forest use, such as sawmills that do not have safety equipment, first aid kits, extinguishers and where personnel lack protection elements.

5.6 Health

As mentioned in the health part of the survey of minors, the environment in which the work carried out is highly dangerous, as well as they can have various accidents, they can contract many diseases, some of these detected only in the long term. In this context, the most frequent diseases of minors are in order of frequency influenza, cuts and wounds, cough, diarrhea, headache. However, a third of the children mentioned that they have not become ill.

5.7 *Share Capital*

The results obtained in this research show that the population centers present in the forest areas where forest use is being carried out is a vegetation consisting mainly of pine and pine-oak forests as well as the so-called scrub and that they are the type of vegetation that covers most of the State. As mentioned above, one of his main focuses of the present research was to determine if the social capital present in the State has contributed to the benefit of those of the forest resources present, as well as to determine the heterogeneity of the nuclei of rural populations through a diagnosis of the state of Nuevo León that defined the levels of organization, development and efficiency in the management of forest resources, and know all kinds of needs. The surveys distinguished the following classes of respondents:

Class I
These people are mainly owners of forest land or holders of these and that at this moment they are not collaborating with any group of people.

Class II
As in class I, this class of people are mainly owners of forest land and that currently if they are collaborating with a group of people and who want to continue collaborating in the same way.

Class III
In this class of people are the owners of forest land and that at this moment they are collaborating with some group of people and who do not want to continue collaborating in the same way.

Forestry activities within the state of Nuevo León have represented a productive sector with technical-productive difficulties due, mainly, to the few areas with a true forestry vocation and the use of silvicultural techniques that turned out not to be the best and most suitable for their conservation and sustainability. Analyzing social capital and its environment that exists in the forest environment of the State, it was possible to find situations that can be related to social capital. It is worth mentioning that the share capital includes all groups of people related to the forestry sector and that in one way or another affect and are affected by the sector. Among them are, therefore, private forest producers, ejido people, community members, as well as providers of forestry technical services, members of the government who have a direct relationship in decision-making regarding the direction that the forestry sector should follow in the entity, academics and researchers, merchants and industrialists or entrepreneurs who carry out operations with the same forestry sector. Among the observed results, it was found that the perception of most respondents (71%), believe that forestry activities have had positive impacts on the state's forestry sector, is reflected in Fig. 1. This shows that the social capital of the State considers that the activities they carry out are good and that they are on the right track, since with them it has been possible to advance in the sector and also that, within the rural area, forest workers are aware of the importance that they play as executing arms of forest

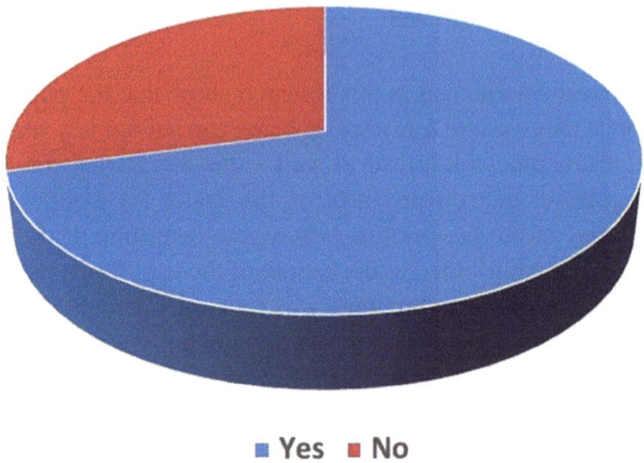

Fig. 1 Response of the villagers to the following question: Do you believe that forestry activities have had positive impacts on the State?

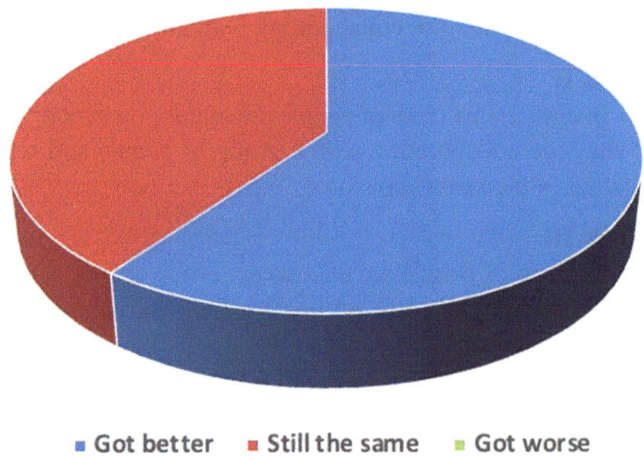

Fig. 2 Behavior of the answers on the relationship of the government with the communities that are found in the forest areas of the state of Nuevo León, according to the perception of the respondents to the question: How do you consider the relationship between the government and the community that works in the forest area of the ejido to which it belongs?

exploitation and when perceiving that their activities have come to benefit the forest resource.

Within the social sector of the forest area, it is considered that social capital does not have very good relations or not the best that would be expected for the sector with the government institutions that interact with the forestry sector and that allows them to develop sustainably and consistently take advantage of the resources available. As can be seen in Fig. 2, 67% of respondents think that relations between forest producers and the government are regular, 16% good and only 17% that think they are very good. It is important to mention that no one thinks in the extremes that the relations between forest producers and government are excellent or bad. Fukuyama (2003), mentions that economic development requires a competent, strong, and effective State, a set of institutions that allow policy change, and the appropriate predisposition of social and political actors. From the above, it can be affirmed the importance of working intensively to improve relations between government and forest producers for the good of the entire forest sector, which combines its potential towards its sustainable development and positively impacts society. Likewise,

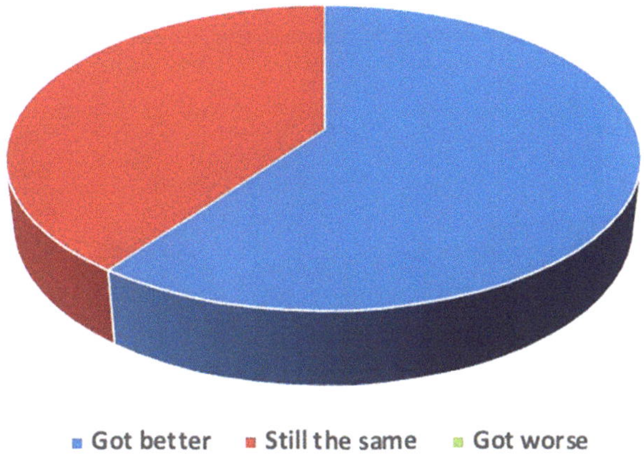

Fig. 3 Respondents' choice regarding the number of forest workers that currently exist in relation to five years ago, regarding the question: Do you think that the number of forest producers in the state of Nuevo León has improved, worst or remains the same?

Bebbington (2003), deals with a reflection of the various trends of environmental and socioeconomic changes, and a reflection of the roles of the different actors of civil society. The focus of the latter refers to peasant organizations in the processes of intensification of rural production, and the use of resources and life strategies. It considers that social capital, which is based on social networks and associative forms of civil society, can have considerable effects on market and government functions. Therefore, it is important that the increase in trust and improvement of social relations between government and producers is generated and leaves both parties and there is the will to do so on both sides.

The perception that forest producers have regarding the increase or reduction of the labor force in these areas found that 50% find that there is the same number of workers, 17% consider that the number is smaller and 33% indicate that this remains the same as can be seen in Fig. 3. These results contrast with the national trend that points to a strong migration from the countryside to the big cities. Therefore, the behavior of the social capital of the forestry sector in the state of Nuevo León is perceived to be increasing in the last 5 years, which is a good expectation for the sector and speaks of the stability of the sector and that the social capital is increasing quantitatively. Knowing that the State is not a purely forestry State, and that it is a productive sector that does not allow the application of the most modern technologies, it is important to know that the sector presents a growth, which according to some of the interviewees, has had a lot to do with the one that from the new century, the forestry sector, not only in the State of Nuevo León, but in the country, has been supported by the federal and state governments, by generating a National Forestry Commission, which is allocated resources to support the country's forestry sector. Hence, the potential social capital in the State is not being lost, but it can become competitive in the future, which depends on its own decisions and these in turn on the integral development of social capital, which goes from its awareness of belonging to the sector, its development of its paradigms to work and live, to the development of organizational strategies that lead social capital to be its own engine of sustainable development that allow ecological and socioeconomic sustainability,

Fig. 4 How to develop activities in the forestry area in the state of Nuevo León. Respondents answer to the question: According to their experience, forest producers, to develop their productive activities, carry them out alone or in groups?

having technology as a tool that transforms the sector through competitiveness and impact on the development of the country. For the forestry sector, organized work in groups is of vital importance, due to the same processes it requires, both in the provision of inputs, tools and technology, as well as for the adequate channeling of its problems and obtaining its solutions, up to marketing activities. Globalization has gradually led the state's forestry sector to work together, the same government programs have in a way promoted it through support programs for the forest sector. This positively impacts the social capital of the state forestry sector because its potential increases exponentially, since the amounts of social capital say little about it, but not the ways of functioning and the synergies that are promoted and exploited, which in turn increases or can increase the competitiveness of the sector and therefore the quality of the state's forest social capital. In Fig. 4 you can see that, in the opinion of the respondents, forest producers in the State carry out their productive activities in groups, the above is said by 67% of the respondents. In the interviews, the above could also be observed. In the area of wildlife management (part of the forestry sector) the tendency is to work in groups, where the participants of the UMA'S support each other, which has generated not only the development of social capital, but also has had a very positive impact on the sustainability of natural resources and therefore forest resources available to the State. This activity of the forestry sector has developed greatly in recent years and has done so in an organized

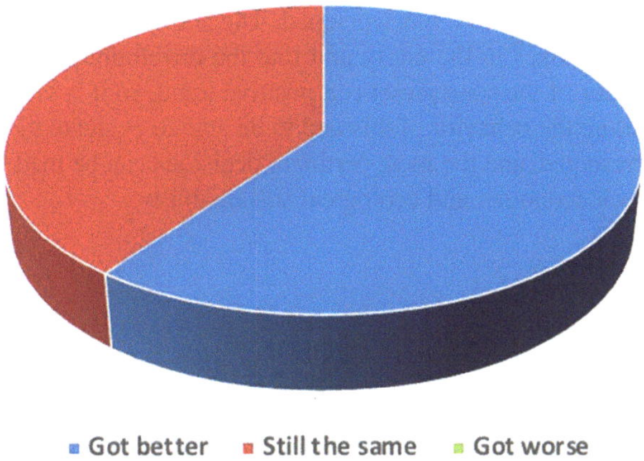

Fig. 5 Consideration of the organized work carried out in the forest productive area in the State. Respondents' answer to the question: You believe that organized work in the forest productive area in the State has become...

way, being an area of the sector where its social capital is being a decisive factor in the development of the forestry sector. As for the causes of why some forest producers continue to work alone, in most of the interviewees and respondents, they agreed that it was a result of their culture and way of thinking, which present distrust between the group or groups of people with whom they interact. Regarding the above, Flora and Flora (2003) define the social capital of ties as the connections between homogeneous individuals and groups, which can be familiar to each other in multiple contexts and therefore, social capital serves as a bridge, which connects various groups within a community with groups outside the community, for that reason the importance of each type of social capital and its relevance for development.

Among these associative resources, they emphasize that they are important to measure the social capital of a group or community, citing the relationships of trust, reciprocity and cooperation, aspects that, as observed in the present research, were aspects that, not presenting themselves among the members of the social capital, did not allow group work and therefore considerably hinder the organization of the forestry sector in the communities. On the new way of working, where it can be observed that forest producers begin to organize themselves to work in groups, it could be observed that it was not a situation that has been easy, nor short-term, but is the result of hard work of forestry technicians and government instances that through activities resulting from planning

Strategically, progress has been made in this regard. Through the transfer of technologies that consider overcoming through demonstrative processes the advantages of the new paradigms to develop forestry activities, this is how progress has been made in this important area to improve the impact of social capital on the sustainable development of the forestry sector in the State (see next photo).

Within the characteristic organization of the forestry sector, which is key to the proper functioning of social capital in the State as a generator of sustainable development, in Fig. 5 it can be clearly seen that most of the interviewees (60%), consider that organized work in the productive area in the State has improved and only 40% think it has worsened. The above speaks of the direction that the forestry sector takes in the entity and that the development of social capital in the forestry sector of the state presents a positive trend, so it is very important to monitor in the future the behavior of this and to be able to evaluate the pace of development that is presented, and the most pertinent decisions can be made regarding the sector and its socioeconomic and ecological sustainability.

6 Discussion

Social capital in the state forestry sector shows a positive trend in terms of the role it plays in the sustainable forest development of the State, so it is very important to monitor in the future the behavior of the state and to be able to evaluate the pace of development that occurs and can make the most relevant decisions regarding the sector and its socioeconomic and ecological sustainability.

The social capital of the forestry sector in the state of Nuevo León has increased in the last 5 years, which is a good expectation for the sector and speaks of the stability of the sector and that the social capital is increasing quantitatively.

The social capital of the forestry sector of the State considers that the activities they carry out are good and that they are on the right track and that with them it has been possible to advance in the sector.

Within the social sector of the forest area, it is considered that social capital does not have very good relations or not the best that would be expected with the governmental bodies that have influence in the sector, which could make it difficult to develop the forest sector in a sustainable way and take advantage of the resources available by social capital in a consistent manner, that allow them to be more efficient and competitive,

The segment of social capital composed of forest producers in the State, tend to carry out their productive activities in groups, which has promoted a change in paradigms to develop the forest sector by increasing the effect of social capital on the sustainable development of the sector. This was observed more advanced in the segment of wildlife management through UMA's and in the solution of complex problems of the forest sector as represented, to mention an example, the control of forest pests. In this new way of working, where it can be observed that forest producers begin to organize themselves to work in groups, it could be observed that it was not a situation that has been easy, nor short-term, but is the result of hard work of forestry technicians and government instances that, through activities resulting from strategic planning, progress has been made in that regard. The potential social capital of the forestry sector in the State is not being lost, but it may become competitive in the future, which depends on its own decisions and these in turn on the integral development of social capital, which goes from its awareness of belonging to the sector, its development of its paradigms to work and live, to the development of organizational strategies that lead social capital to be its own engine of sustainable development, which allow ecological and socioeconomic sustainability, having technology as a tool that transforms the sector through competitiveness and impact on the development of the country. The perception of the inhabitants of rural areas with forest resources about education is highly positive. It was found that 85% of forest producers can read and write. The nuclear family is the hegemonic among the population of the state's forest sector. 88 percent reported living in this type of family, only 5 percent in families headed by the mother and the remaining 7 percent with an extended family (grandparents, uncles, mainly). The study allowed to distinguish three classes of informants in the forestry sector of the State: Class I, these people are mainly owners of forest land or holders of these and that at this moment they are not collaborating with any group of people. Class II, as in class I, this class of people are mainly owners of forest land or holders of them and that at this time if they are collaborating with a group of people and who wish to continue collaborating in the same way and Class III, in this class of people are the owners of forest land or holders of them and who at this time are collaborating with some group of people who do not want to continue collaborating in the same way.

References

Aguirre O (2015) Manejo Forestal en el Siglo XXI. Madera Bosques 21(Núm Esp):17–28

Bebbington A (2003) Capital social y reducción de la pobreza en América Latina y el Caribe: en busca de un nuevo paradigma. Capital social e intensificación de las estrategias de vida: organizaciones locales e islas de sostenibilidad en los Andes rurales. pp 491–508

Conagua (2018) Reporte del clima en México. Coordinación General del Servicio Meteorológico Nacional Gerencia de Meteorología y Climatología Reporte del Clima en México Subgerencia de Pronóstico a Mediano y Largo Plazo. Año 8 Número 4. 32pp

Diaz J, Albertini F (2003) Capital social y reducción de la pobreza en América Latina y el Caribe: en busca de un nuevo paradigma. Organizaciones de base y el Estado: recuperando los eslabones perdidos de la sociabilidad. pp 247–302

Durston J (2003) Capital social y reducción de la pobreza en América Latina y el Caribe: en busca de un nuevo paradigma. Parte del problema, parte de la solución, su papel en la persistencia y en la superación de la pobreza en América Latina y el Caribe. pp 147–202

Flora J, Flora C (2003) Capital social y reducción de la pobreza en América Latina y el Caribe: en busca de un nuevo paradigma. Desarrollo comunitario en las zonas rurales de los Andes. pp 555–580

Flores M, Rello F (2003) Capital social y reducción de la pobreza en América Latina y el Caribe: en busca de un nuevo paradigma. Capital social: virtudes y limitaciones. pp 203–228

Fukuyama F (2003 Capital social y reducción de la pobreza en América Latina y el Caribe: en busca de un nuevo paradigma. Capital social y desarrollo: la agenda venidera. pp 33–50

García J (2011) Una definición estructural de capital social. Department of Sociology. University of Leicester. REDES- Revista hispana para el análisis de redes sociales 20(6):132–160

Instituto Nacional de Estadística, Geografía e Informática (INEGI) (2019) Mapas. Área geográfica. https://www.inegi.org.mx/app/mapas/. Consultado el 24 de mayo de 2019

Lutz B (2011) El capital social en discusión. Caso del desarrollo rural en México. El capital social en discusión. pp 69–93

Martínez D, Reyes J, Andrés A, Morales G (2015) Uso y manejo de los recursos maderables como combustible en la comunidad de San Pedro Alpatláhuac, Cuauhtinchán, Puebla, México. Revista Iberoamericana de Ciencias ISSN 2334-2501. 2(1). 17pp

Observatorio Social del Agro Mesoamericano (OSAM) (2009) Las comunidades forestales y el aprovechamiento –conservación de los recursos naturales
Seminis (2018) El Gran Potencial De La Industria Agrícola Mexicana. https://www.seminis.mx/el-gran-potencial-de-la-industria-agricola-mexicana/. Consultado el 24 de mayo de 2019
Robison L, Siles M, Schmid A (2003) Capital social y reducción de la pobreza en América Latina y el Caribe: en busca de un nuevo paradigma. Hacia un paradigma maduro. pp 51–114
Rojas R (1998) Guía para realizar investigaciones sociales. Vigésima edición, Editorial Plaza y Valdés. p 437
Semarnat (2012) Informe de la Situación del Medio Ambiente en México. Compendio de Estadísticas Ambientales. Indicadores Clave y de Desempeño Ambiental. Edición 2012. México. 360pp
Sunkel G (2003) Capital social y reducción de la pobreza en América Latina y el Caribe: en busca de un nuevo paradigma. La pobreza en la ciudad: capital social y políticas públicas. pp 303–338
Treviño E (2009) Reversión de los procesos de desertificación mediante técnicas agroforestales en Nuevo León
Uphoff N (2003) Capital social y reducción de la pobreza en América Latina y el Caribe: en busca de un nuevo paradigma. El capital social y su capacidad de reducción de la pobreza. pp 115–146
Urteaga E (2013) La teoría del capital social de Robert Putnam: Originalidad y carencias. Universidad Autónoma de Bucaramanga Bucaramanga, Colombia. Reflexión Política, 15(29): 44–60

Effect of High Temperatures That Simulate Climate Change in the Germination of Seven Species of the Tamaulipan Thornscrub

Regina Pérez-Domínguez and Wibke Himmelsbach

Abstract The increase in temperature and changes in precipitation patterns, affect all ecosystems and has a great influence on plant species populations. The aim of this investigation is to evaluate the germination capacity and 50% germination under controlled conditions of temperature and photoperiod of seven species of Tamaulipan Thornscrub simulating the temperature increase in a possible climate change. With the results generated, we can predict which species has the capacity to germinate at different temperatures increments, which species can adapt to climate change, and which will migrated or will become extinct. *Caesalpinia mexicana, Celtis pallida, Condalia hookeri, Ehretia anacua, Parkinsonia aculeata, Prosopis glandulosa* and *Prosopis laevigata* were studied. Each specie. counted with seeds from different populations, which were scarified, seeded, and placed in the incubator at different temperatures (average temperature, average temperature +2°C, average temperature +5°C and average temperature +7°C). The species that were able to germinate in the four treatments were: *C. mexicana, P. aculeata, P. glandulosa* and *P. laevigata*, whereas *C. pallida, C. hookeri* and *E. anacua* did not germinate in 7°C above the average temperature. The results show that the studied species of Tamaulipan Thornscrub have the capacity to adapt to the temperatures increment as an effect of climate change.

Keywords Climate change · Germination capacity · T50 · Temperature · Tamaulipan thornscrub

1 Introduction

This study evaluates the germination capacity and germination rate in controlled conditions of photoperiod and temperature of 7 species of the Tamaulipan Thornscrub of different origins under 4 different temperatures that simulate an

R. Pérez-Domínguez (✉) · W. Himmelsbach
Facultad de Ciencias Forestales, Universidad Autónoma de Nuevo León, Linares, Nuevo León, Mexico
e-mail: regina.perezdm@uanl.edu.mx

F. Garza-Ocañas (ed.), *Sustainable Management of Natural Resources*, Earth and Environmental Sciences Library, https://doi.org/10.1007/978-3-031-33394-1_7

increase in temperature in the face of possible climate change. In recent years, climate change has been associated with an increase in temperatures and a change in precipitation in its various forms due to an increase in greenhouse gases (Arnell et al. 2019; IPCC 2007). The increase in greenhouse gases and temperatures affect all ecosystems in the world (Evangelista et al. 2016; Solomon et al. 2007) and have a great influence on plant species populations (Herranz et al. 2009; Addler and Hillie Ris Lambers 2008; Nicotra et al. 2010). The effects can result in changes in phenology, species distribution and can cause species extinction (Inouye 2020; Easterling et al. 2000). High temperatures cause a reduction in biomass in plant species, due to the decrease in photosynthesis and the increase in transpiration and stomatal conduction (Jones 1992), mainly affecting the progeny of plants. An increase in aridity and temperature can on the one hand decrease the presence of some species, but it can also promote others to adapt to changes (Peñuelas and Filella 2001). Germination is known to be a complex process in which genetic interactions and environmental factors are related (Basbag et al. 2009). One of the main physical factors that affect germination after water availability is temperature, because it affects the percentage and speed of germination, as well as the dormancy or inhibition of germination (Baskin and Baskin 1998; Shimono and Kudo 2005). Some plant species can germinate in a wide range of temperatures, while others only do so in a very narrow range (Lai et al. 2016, Wang et al. 2010, Probert 2000 in Fenner 2000, Fenner 1985). Similarly, seeds of the same species but from different origins may require different temperatures to germinate due to the environment in which they develop and their phenotypic plasticity (Lai et al. 2016; Cavieres and Arroyo 2000). It has been shown that carbon dioxide concentration, oxygen concentration, temperature, and availability of water and nutrients can affect seeds in their germination and emergence (Borchet et al. 1989). The success of seedling emergence depends primarily on the seed's ability to germinate and the ability to emerge to the soil surface (Aou-ouad et al. 2014). Several studies show that the survival of seedlings during the early stages is a critical phase for their establishment and success of plant species (Grossnickle and MacDonald 2018; Fenner and Kitajima 1999).

2 Methodology

2.1 Selection and Treatment of Plant Species

The species of the Tamaulipan Thornscrub selected for this study have a high abundance in Mexico, and are of great economic importance (firewood, fence posts, construction of houses, manufacture of furniture, fodder, etc.) (Estrada Castillon et al. 2004; Reid et al. 1990) and ecological (rehabilitation of degraded areas and ecological value). The species selected for this work were: *Caesalpinia mexicana* Gray (arboreal), *Celtis pallida* Torr. (shrubby), *Condalia hookeri* M. C. Jhonst. (shrubby), *Ehretia anacua* I. M. Jhonst. (arboreal), *Parkinsonia aculeata*

Table 1 Information on the origins of the species studied

Species	Municipality/ Community	State	Altitude (msnm)	Latitude N	Longitude O
Caesalpinia mexicana Gra	Linares	NL	370	24°47'45"	99°32'31"
Caesalpinia mexicana Gra	Ejido Rancho Viejo y la Pal	NL	710	24°44'57"	99°47'14"
Celtis pallida Torr.	Linares	NL	370	24°47'45"	99°32'31"
Celtis pallida Torr.	Carretera Linares-Iturbide	NL	450	24°47'15"	99°39'13"
Celtis pallida Torr.	Ejido Los Ángeles	NL	560	24°57'37"	99°49'20"
Condalia hookeri M. C. Jh	Linares	NL	370	24°47'45"	99°32'31"
Condalia hookeri M. C. Jho	Carretera Linares-Iturbide	NL	450	24°47'15"	99°39'13"
Ehretia anacua I. M. Jhons	Linares	NL	370	24°47'45"	99°32'31"
Ehretia anacua I. M. Jhons	Ejido Los Ángeles	NL	560	24°57'37"	99°49'20"
Parkinsonia aculeata L.	Linares	NL	370	24°47'45"	99°32'31"
Parkinsonia aculeata L.	Ejido Los Ángeles	NL	560	24°57'37"	99°49'20"
Parkinsonia aculeata L.	Monclova	Coah.	650	26°52'27"	101°25'22
Parkinsonia aculeata L.	Torreón	Coah.	1215	25°31'27"	103°274'1
Parkinsonia aculeata L	San Luis Potosí	SLP	1890	22°07'52"	100°59'37
Prosopis glandulosa Torr.	Galeana	NL	1630	24°49'04"	100°04'20
Prosopis glandulosa Torr.	Ejido Llanos de la Unión	Coah.	1987	25°23'52"	101°07'24
Prosopis glandulosa Torr.	Ejido Angostura	Coah.	1785	25°20'32"	101°02'42
Prosopis laevigata Willd.	Linares	NL	370	24°47'45"	99°32'31"
Prosopis laevigata Willd.	Ejido La Soledad	NL	1581	24°00'18"	100°03'27
Prosopis laevigata Willd.	San Luis Potosí	SLP	1890	22°07'52"	100°59'37

L. (arboreal), *Prosopis glandulosa* Torr., (arboreal) and *Prosopis laevigata* Willd. (arboreal).

Seed collection was carried out during spring-summer 2011 and 2012 in the states of Nuevo León, San Luis Potosí, and Coahuila (Table 1), at different elevations. For each species and origin, seeds were collected from at least 10 mother plants to include local genetic variation.

Once the seeds were collected, those that were free of damage and that presented good quality were selected, to later scarify them with sandpaper, manually. Once this procedure was carried out, the plastic trays were prepared with (288 cavities, 10.29 mL/cavity) mountain soil, peat moss and vermiculite in the proportions 6: 3: 1 seed was sown per cavity for the case of large seeds, 2 seeds for medium size and 4 for small ones. For each species/origin, four blocks were made with five repetitions each.

2.2 *Evaluation of the Percentage and Speed of Germination*

To determine the percentage and speed of germination (T50) of the species under study, a Lumistell Germinator (laboratory) was used, to simulate the increase in temperature because of climate change. Soil temperatures were taken as a basis in the first 5 cm of depth of the month of September, which is the rainiest month and in which most species germinate (Jurado et al. 2006; Jurado et al. 2000), from the closest weather station to the area (www.wcc.nrcs.usda.gov). First, daytime (7: 00am–5:00pm) and nighttime (6:00pm–6:00am) temperatures were considered with one-hour intervals, remaining constant from 6:00 pm to 6:00 am. Subsequently, the hour-by-hour increments of the period from 7:00 am to 5: 00 p.m. were obtained, to simulate what happens as close to reality as possible. The experiment was based on three treatments and a control, the temperatures used were the most pessimistic about climate change and were: real temperature of the month of September 2011 (Witness), real temperature plus 2 °C (T1), real temperature plus 5 °C (T2) and real temperature plus 7 °C (T3) (Appendix 1), with 12 light hours and 12 hours of darkness. The germination rate (T50) was determined by the number of days in which 50% of the seeds emerged (Bewley and Black 1982). The percentage of germination was determined, counting the seeds that emerged each day for 15 days and at the end of the experiment the following formula was used:

$$\%G = 100^{*}N/NS$$

Where: N: Number of seeds that emerged NS: Total number of seeds

2.3 *Statistical Analysis*

For the percentages of germination and germination speed in the four treatments, they were statistically analyzed to determine the differences between the origins, the temperatures and their possible interaction between the temperatures and origins, by means of a nested factorial analysis of variance model. When differences between treatments were observed ($P < 0.05$) Tukey mean comparison tests were applied. This analysis was performed with the statistical software "R Project". The statistical model used was as follows:

$$Y_{ijk} = \mu + \beta_i + Ti + \beta P_{ij} + A(P)_{kj} + E_{ijk}$$

Where: Yijk = observed value of the k-th family of the j-th origin, at the i-th temperature; μ = population mean; βi = random effect of the i-th origin; Pj = random effect of the j-th temperature; βPij = random effect of the interaction of the i-th origin with the j-th temperature; A(P)kj = random effect of the k-th family nested in the j-th origin; Eijk = random sampling error.

3 Results

3.1 Germination Percentage

The Fig. 1 shows that *Caesalpinia mexicana. Condalia hookeri, Celtis pallida* and *Ehretia anacua* behaved in the same way, regardless of temperature and origin, they germinated less at higher temperatures in all their origins. *Caesalpinia mexicana, Parkinsonia aculeata, Prosopis glandulosa, Prosopis laevigata* are species that germinate in all temperatures, although their behavior is different between treatments (Fig. 1). Seeds of *Celtis pallida* and *Condalia hookeri* showed a similar behavior. That is, a similar germination was observed between origins (*Celtis pallida*, F = 1.402, g.l. = 2, P = 0.259; *Condalia hookeri*, F = 3.333, gl. = 1, P = 0.085), but different in treatments (*Celtis pallida*, F = 17.958, g.l. = 3, P < 0.001; *Condalia hookeri*, F = 7.467, gl. = 3 P < 0.001). As the soil temperature increased, the percentage of germination was lower in both species and origins. The interaction between origins and treatments were not significant (*Celtis pallida*, F = 0.444, g.l. = 6, P = 0.844; *Condalia hookeri*, F = 0.200, gl. = 3, P = 0.331). *Ehretia anacua, Parkinsonia aculeata, Prosopis glandulosa* and *P. laevigata* behaved in the same way, that is, they presented a difference in the percentage of germination between the origins (*Ehretia anacua*: F = 4.417, g.l. = 1, P = 0.046; *Parkinsonia aculeata*: F = 3.487, g.l. = 4, P = 0.012; *Prosopis glandulose*: F = 5.874, g.l. = 2, P = 0.006 and *Prosopis laevigata*: F = 3.783, g.l. = 2, P = 0.032) and between treatments (*Ehretia anacua*: F = 7.617, g.l. = 3, P < 0.001; *Parkinsonia aculeata*: F = 10.745, g.l. = 3, P < 0.001; *Prosopis glandulose*: F = 2.997, g.l. = 3, P = 0.043 and *P.laevigata*: F = 5.926, g.l. = 3, P = 0.002). Likewise, the interaction between the origins and treatments was significant for (*Ehretia anacua*: F = 11.108, g.l. = 3, P < 0.001; *Parkinsonia aculeata*: F = 2.365, g.l. = 12, P = 0.014; *Prosopis glandulosa*: F = 3.021, g.l. = 6, P = 0.017 and *P. laevigata*: F = 3.559, g.l. = 6, P = 0.007) (Table 2).

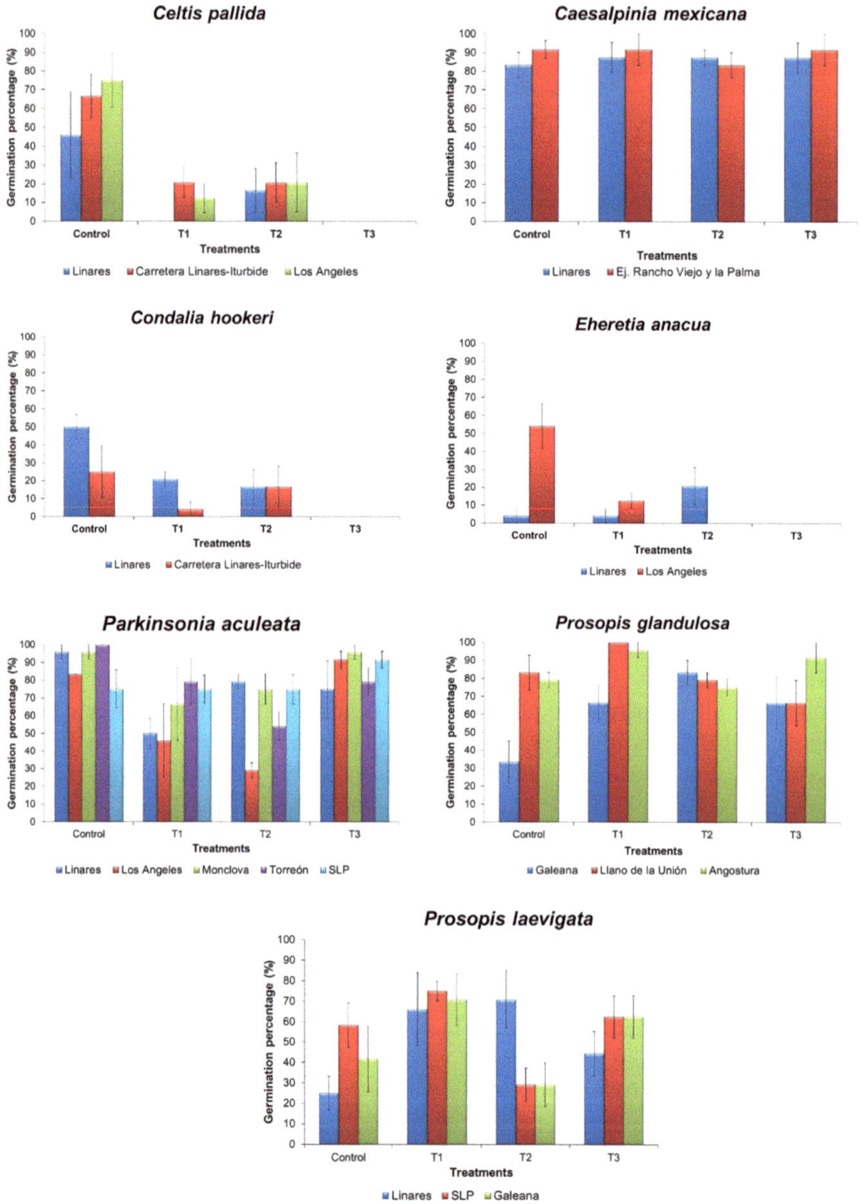

Fig. 1 Effect of seed provenance on germination percentage (Average ± E.S) in four treatments (control = average temperature, T1 = average temperature + 2, T2 = average temperature + 5, T3 = average temperature + 7)

Table 2 Germination percentages showing significant differences (p < 0.05) with different lowercase letters between temperatures, origins, and interactions (I) between origin: temperature

Information of the especies	Temperature for germination Procedence									
Especie	Proc	Testigo	I	T1	I	T2	I	T3	I	
Caesalpinia mexicana	1	83 ± 13.61	a	88 ± 15.96	a	88 ± 8.33	a	88 ± 15.96	a	a
	2	92 ± 9.62	a	92 ± 16.67	a	83 ± 13.61	a	92 ± 16.67	a	a
Temperature		a	a	a	a					
Celtis pallida	1	46 ± 45.90	a	0 ± 0	b	17 ± 23.57	ab	0 ± 0	b	a
	3	67 ± 23.57	a	21 ± 15.96	ab	21 ± 20.97	ab	0 ± 0	b	a
	4	78 ± 28.87	a	13 ± 15.96	ab	21 ± 31.55	ab	0 ± 0	b	a
Temperature		a	b	b	b					
Condalia hookeri	1	50 ± 13.61	a	21 ± 8.33	ab	17 ± 19.25	ab	0 ± 0	b	a
	3	25 ± 28.87	ab	4 ± 8.33	b	17 ± 23.57	ab	0 ± 0	b	a
Temperature		b	a	b	a					
Ehretia anacua	1	4 ± 8.34	b	4 ± 8.33	b	21 ± 20.97	a	0 ± 0	b	a
	4	54 ± 25.00	a	13 ± 8.34	a	0 ± 0	b	0 ± 0	b	b
Temperature		b	a	a	a					
Parkinsonia aculeata	1	96 ± 8.33	a	58 ± 16.67	a	79 ± 8.33	ab	75 ± 31.91	ab	ab
	4	83 ± 0	a	38 ± 41.67	b	28 ± 8.33	b	92 ± 9.62	a	b
	5	96 ± 8.33	a	67 ± 40.82	a	75 ± 16.67	ab	96 ± 8.33	a	a
	6	100 ± 0	a	88 ± 25.00	a	54 ± 15.96	ab	79 ± 15.96	ab	a
	7	75 ± 21.52	a	79 ± 15.96	a	75 ± 16.67	ab	92 ± 9.62	a	a
Temperature		b	a	a	b					
Prosopis glandulosa	8	33 ± 23.57	b	67 ± 19.25	a	83 ± 13.61	a	71 ± 28.46	ab	ab
	9	83 ± 19.24	a	100 ± 0	a	79 ± 8.33	a	62 ± 25	ab	a
	10	79 ± 8.28	a	96 ± 8.34	a	75 ± 9.61	ab	92 ± 16.66	a	b
Temperature		b	a	ab	ab					
Prosopis laevigata	1	25 ± 16.50	b	67 ± 36.00	a	63 ± 28.48	a	42 ± 21.52	ab	ab
	8	42 ± 32.12	ab	71 ± 25.00	a	75 ± 21.52	a	63 ± 20.97	a	a
	7	58 ± 21.56		75 ± 9.62		29 ± 15.96	ab	63 ± 20.97	a	b
Temperature		b	a	ab	b					

Results show that *Caesalpinia mexicana* was the only species responding in the same way in the four conditions of temperature-origin ($F = 0.392$, g.l.=1, $P = 0.531$) and treatments ($F = 0.160$, g.l. =3, $P = 0.0922$)

Abbreviations: Pro = Provenance: 1= Linares; 2= Ejido Rancho Viejo; 3 = Carretera Linares-Iturbide; 4 = Ejido Los Ángeles; 5 = Monclova Coahuila; 6 = Torreón Coahuila; 7 = San Luis Potosi; 8 = Galeana; 9 = Llano de la Unión; 10 = Angostura

3.2 Germination Rate

Table 3 shows the germination rates of seven species, as well as their level of significance between the treatments (temperatures), origins of the seeds and their interaction. Where it was observed that *Caesalpinia mexicana, Celtis pallida, Condalia hookeri, Ehretia anacua Parkinsonia aculeata* and *Prosopis laevigata* did not present differences in their germination rate between the origins, but in the treatments. *P. glandulosa* presented differences between its origins (F = 5.16, g.l. = 2, P = 0.010) and between its treatments (F = 66.04, g.l. = 3, P < 0.001). *P. glandulosa* and *P. laevigata* were the species that germinated in the shortest

Table 3 Germination rate of each of the species and their origins and the significant differences (p < 0.05) are shown with different lowercase letters (a, b, c) between temperatures and origins

Species	Provenance	Control	T1	T2	T3	Provenance
Caesalpinia mexicana	Linares	4	3	4	4	3.8^{a}
	Ejido Rancho Viejo y La Palm	3.3	3	4.8	3.5	3.7^{a}
Temperature		3.6^{a}	2.7^{b}	4.3^{c}	3.2^{a}	
Celtis pallida	Linares	9.3	–	10	–	9.7^{a}
	Carretera Linares-Iturbide	9.5	7.7	8.3	–	8.5^{a}
	Ejido Los Ángeles	8.8	6	8.5	–	7.8^{a}
Temperature		9.2^{b}	6.9^{a}	8.9^{ab}	–	
Condalia hookerii	Linares	11	6.5	9.5	–	9.0^{a}
	Carretera Linares-Iturbide	11.8	9	10	–	10.3^{a}
Temperature		8.8^{b}	6.1^{a}	7.9^{a}	–	
Ehretia anacua	Linares	11	11	10	–	10.7^{a}
	Ejido Los Ángeles	12	7	–	–	9.5^{a}
Elevaciones		10.6^{a}	8.0^{a}	9.0^{a}	$–^{b}$	
Parkinsonia aculeata	Linares	4.5	2.5	4.3	7.3	4.7^{a}
	Ejido Los Ángeles	4.8	3.5	4.3	6.5	4.8^{a}
	Monclova, Coahuila	4.8	3.5	4.4	7	4.9^{a}
	Torreón, Coahuila	4.3	3	4.5	6	4.5^{a}
	San Luis Potosí	5.3	2.6	6	5	4.7^{a}
Temperature		4.8^{a}	3.0^{b}	5.0^{a}	6.0^{c}	
Prosopis glandulosa	Galeana	4.3	2.3	4	2	3.2^{a}
	La Soledad	2.8	2	4	2	2.7^{b}
	Llano de la Unión	3	2	4	2	2.8^{b}
Temperature		3.4^{b}	2.1^{a}	4.0^{b}	2.0^{a}	
Prosopis laevigata	Linares	2.3	2.5	3	3.8	2.9^{a}
	Galeana, N.L.	3	2	4	2	2.8^{a}
	San Luis Potosí	3.5	2	3.5	2	2.8^{a}
Temperature		2.9^{ab}	2.2^{a}	3.5^{b}	2.6^{a}	

time, meaning that 50 percent of the seeds germinated on average in 2.8 and 2.9 days. While *C. hookeri* (9.7 days) and *E. anacua* (10.1 days) germinated in a longer time.

4 Discussion

The difference in germination capacity between different populations of the same species is interpreted as an adaptation to the characteristics of the environment (climatic factors) from which they come (Meyer et al. 1997). In the present study, all the species were able to germinate in the 3 treatments carried out, except for *C. pallida, C. hookeri* and *E. anacua*, which failed to germinate at high temperatures (T3 = real temperature + 5 ° C), so it can be predicted that these species will present difficulties in their germination capacity in case the temperature increased more than 5 °C. According to Villers-Ruiz and Trejo-Vázquez (1997), scrub vegetation is one of the ecosystems that can be sensitive to climate change according to the scenarios presented, presenting some changes in its structure. The high variation between seed origins and treatments (temperatures) (*E. anacua, P. aculeata, P. glandulosa* and *P. laevigata*), is perhaps due to the interaction between genotypes and environmental conditions (Vassilevska-Ivanova and Tcekova 2002). The seeds of *C. mexicana* and *P. aculeata* had a high percentage of germination in treatment 3 (T3 = real temperature + 7°C). With the results obtained in this study, it can be predicted that all the species studied can germinate at high temperatures and that they will be able to adapt to climate change.

Appendix

Table 4 Actual soil temperatures occurring in the month of September 2011

Days/Normal temperature															
Time	*1*	*2*	*3*	*4*	*5*	*6*	*7*	*8*	*9*	*10*	*11*	*12*	*13*	*14*	*15*
07:00	28.8	30.1	28.8	30.1	31.3	32.6	32.6	32.6	32.6	31.3	30.1	31.3	30.1	30.1	30.1
08:00	29.0	30.3	29.0	30.3	31.5	32.7	32.7	32.7	32.8	31.5	30.3	31.5	30.3	30.3	30.3
09:00	29.5	30.6	29.5	30.6	31.8	32.9	32.9	32.9	33.3	32.0	30.8	32.0	30.8	30.8	30.6
10:00	30.4	31.0	30.1	31.0	32.2	33.1	33.1	33.1	33.8	32.6	31.7	32.7	31.5	31.7	30.9
11:00	31.7	31.4	30.7	31.4	32.6	33.3	33.3	33.3	34.3	33.2	33.0	33.8	32.6	33.0	31.2
12:00	33.2	31.7	31.4	31.7	32.9	33.5	33.5	33.5	34.9	33.9	34.3	34.9	33.7	34.3	31.5
13:00	34.4	32.1	32.0	32.1	33.2	33.6	33.6	33.6	35.4	34.5	35.5	36.0	34.8	35.5	31.8
14:00	35.3	32.5	32.6	32.5	33.6	33.7	33.7	33.7	35.6	34.7	36.5	36.8	35.6	36.5	32.0
15:00	35.8	32.8	32.8	32.8	33.8	33.8	33.8	34.8	35.8	33.8	35.8	35.8	34.8	35.8	30.8
16:00	35.6	32.6	32.6	32.6	33.6	33.6	33.6	34.6	35.6	33.6	35.6	35.6	34.6	35.6	30.6
17:00	30.6	27.6	27.6	27.6	28.6	28.6	28.6	29.6	30.6	28.6	30.6	30.6	29.6	30.6	25.6

Table 5 Actual soil temperatures occurring in the month of September 2011 +2 °C

Days/Normal temperature +2															
Time	*1*	*2*	*3*	*4*	*5*	*6*	*7*	*8*	*9*	*10*	*11*	*12*	*13*	*14*	*15*
07:00	30.8	32.1	30.8	32.1	33.3	34.6	34.6	34.6	34.6	33.3	32.1	33.3	32.1	32.1	32.1
08:00	31.0	32.3	31.0	32.3	33.5	34.7	34.7	34.7	34.8	33.5	32.3	33.5	32.3	32.3	32.3
09:00	31.5	32.6	31.5	32.6	33.8	34.9	34.9	34.9	35.3	34.0	32.8	34.0	32.8	32.8	32.6
10:00	32.4	33.0	32.1	33.0	34.2	35.1	35.1	35.1	35.8	34.6	33.7	34.7	33.5	33.7	32.9
11:00	33.7	33.4	32.7	33.4	34.6	35.3	35.3	35.3	36.3	35.2	35.0	35.8	34.6	35.0	33.2
12:00	35.2	33.7	33.4	33.7	34.9	35.5	35.5	35.5	36.9	35.9	36.3	36.9	35.7	36.3	33.5
13:00	36.4	34.1	34.0	34.1	35.2	35.6	35.6	35.6	37.4	36.5	37.5	38.0	36.8	37.5	33.8
14:00	37.3	34.5	34.6	34.5	35.6	35.7	35.7	35.7	37.6	36.7	38.5	38.8	37.6	38.5	34.0
15:00	37.8	34.8	34.8	34.8	35.8	35.8	35.8	36.8	37.8	35.8	37.8	37.8	36.8	37.8	32.8
16:00	37.6	34.6	34.6	34.6	35.6	35.6	35.6	36.6	37.6	35.6	37.6	37.6	36.6	37.6	32.6
17:00	32.6	29.6	29.6	29.6	30.6	30.6	30.6	31.6	32.6	30.6	32.6	32.6	31.6	32.6	27.6

Table 6 Actual soil temperatures occurring in the month of September 2011 +5° C

Days/Normal temperature+5															
Time	*1*	2	3	*4*	5	*6*	7	*8*	*9*	*10*	*11*	*12*	*13*	*14*	*15*
07:00	33.8	35.1	33.8	35.1	36.3	37.6	37.6	37.6	37.6	36.3	35.1	36.3	35.1	35.1	35.1
08:00	34.0	35.3	34.0	35.3	36.5	37.7	37.7	37.7	37.8	36.5	35.3	36.5	35.3	35.3	35.3
09:00	34.5	35.6	34.5	35.6	36.8	37.9	37.9	37.9	38.3	37.0	35.8	37.0	35.8	35.8	35.6
10:00	35.4	36.0	35.1	36.0	37.2	38.1	38.1	38.1	38.8	37.6	36.7	37.7	36.5	36.7	35.9
11:00	36.7	36.4	35.7	36.4	37.6	38.3	38.3	38.3	39.3	38.2	38.0	38.8	37.6	38.0	36.2
12:00	38.2	36.7	36.4	36.7	37.9	38.5	38.5	38.5	39.9	38.9	39.3	39.9	38.7	39.3	36.5
13:00	39.4	37.1	37.0	37.1	38.2	38.6	38.6	38.6	40.4	39.5	40.5	41.0	39.8	40.5	36.8
14:00	40.3	37.5	37.6	37.5	38.6	38.7	38.7	38.7	40.6	39.7	41.5	41.8	40.6	41.5	37.0
15:00	40.8	37.8	37.8	37.8	38.8	38.8	38.8	39.8	40.8	38.8	40.8	40.8	39.8	40.8	35.8
16:00	40.6	37.6	37.6	37.6	38.6	38.6	38.6	39.6	40.6	38.6	40.6	40.6	39.6	40.6	35.6
17:00	35.6	32.6	32.6	32.6	33.6	33.6	33.6	34.6	35.6	33.6	35.6	35.6	34.6	35.6	30.6

Table 7 Actual soil temperatures occurring in the month of September 2011 +7 °C

Days/Normal temperature+7															
Time	*1*	2	*3*	*4*	*5*	*6*	*7*	*8*	*9*	*10*	*11*	*12*	*13*	*14*	*15*
07:00	35.8	37.1	35.8	37.1	38.3	39.6	39.6	39.6	39.6	38.3	37.1	38.3	37.1	37.1	37.1
08:00	36.0	37.3	36.0	37.3	38.5	39.7	39.7	39.7	39.8	38.5	37.3	38.5	37.3	37.3	37.3
09:00	36.5	37.6	36.5	37.6	38.8	39.9	39.9	39.9	40.3	39.0	37.8	39.0	37.8	37.8	37.6
10:00	37.4	38.0	37.1	38.0	39.2	40.1	40.1	40.1	40.8	39.6	38.7	39.7	38.5	38.7	37.9
11:00	38.7	38.4	37.7	38.4	39.6	40.3	40.3	40.3	41.3	40.2	40.0	40.8	39.6	40.0	38.2
12:00	40.2	38.7	38.4	38.7	39.9	40.5	40.5	40.5	41.9	40.9	41.3	41.9	40.7	41.3	38.5
13:00	41.4	39.1	39.0	39.1	40.2	40.6	40.6	40.6	42.4	41.5	42.5	43.0	41.8	42.5	38.8
14:00	42.3	39.5	39.6	39.5	40.6	40.7	40.7	40.7	42.6	41.7	43.5	43.8	42.6	43.5	39.0
15:00	42.8	39.8	39.8	39.8	40.8	40.8	40.8	41.8	42.8	40.8	42.8	42.8	41.8	42.8	37.8
16:00	42.6	39.6	39.6	39.6	40.6	40.6	40.6	41.6	42.6	40.6	42.6	42.6	41.6	42.6	37.6
17:00	37.6	34.6	34.6	34.6	35.6	35.6	35.6	36.6	37.6	35.6	37.6	37.6	36.6	37.6	32.6

References

Addler PB, Hillie Ris Lambers J (2008) The influence of climate and species composition on the population dynamics of ten prairie forbs. Ecology 89:3049–3060

Aou-ouad H, Medrano H, Lamarti A, Gulías J (2014) Seed germination at different temperatures and seedling emergence at different depths of *Rhamnus* spp. Cent Eur J Biol 9:569–578

Arnell NW, Lowe JA, Challinor AJ, Osborn TJ (2019) Global and regional impacts of climate change at different levels of global temperature increase. Clim Change 155:377–391

Basbag M, Toncer O, Basbag S (2009) Effects of different temperaturas and duration on germination of caper (*Capparis ovate*) seeds. J Environ Biol 30(4):621–624

Baskin CC, Baskin JM (1998) Seeds. Ecology, biogeography, and evolution of dormancy and germination. Academic Press, San Diego

Bewley JD, Black M (1982) Physiology and biochemistry of seeds in relation to germination, Vol. II. Viability, dormancy and environmental control. Springer-Verlag, Berlin

Borchet MI, Davis FW, Michaelsen J, Oyler LD (1989) Interactions of factor affecting seedling recruitment of blue oak (*Quercus douglasii*) in California. Ecology 70:389–404

Cavieres LA, Arroyo MTK (2000) Seed germination response to cold stratification period and thermal regime in Phacelia secunda (Hydrophyllaceae) – altitudinal variation in the Mediterranean Andes of central Chile. Plant Ecol 149:1–8

Easterling DR, Meehl GA, Parmesan C, Changnon SA, Karl TR, Mearns LO (2000) Climate extremes: observations, modeling, and impacts. Science 289:2068–2074

Estrada Castillon E, Yen Mendes C, Delgado A, Villarreal Quintanilla JA (2004) Leguminosas del centro del estado de Nuevo León, México. Anales del Instituto de Biología, UNAM. Serie Botánica 75(1):73–85

Evangelista A, Frate L, Carranza ML, Attorre F, Pelino G, Stanisci A (2016) Changes in composition, ecology, and structure of high-mountain vegetation: a re-visitation study over 42 years. AoB Plants 8:plw004. https://doi.org/10.1093/aobpla/plw004

Fenner M (1985) Seed ecology. Chapman and Hall, p 149

Fenner M (2000) Seeds: the ecology of regeneration in plant communities. Caby Publishing, p 413

Fenner M, Kitajima K (1999) Seed and seedling ecology. In: Pugnaire FI, Valladares F (eds) Handbook of functional plant ecology. Marcel Dekker, New York, pp 589–611

Grossnickle SC, MacDonald JE (2018) Seedling quality: history, application, and plant attributes. Forests 9(5):283

Herranz JM, Copete MA, Ferrandis P (2009) Posibles efectos del cambio climático sobre las especies vegetales en Castilla-La Mancha. Impactos del cambio climático Castilla-La Mancha. Ed. Fundación General del Medio Ambiente. pp 203–317

Inouye DW (2020) Effects of climate change on alpine plants and their pollinators. Ann N Y Acad Sci 1469:26–37. https://doi.org/10.1111/nyas.14104

IPCC-WGI (2007). Intergovernmental Panel on Climate Change, Working Group I. Working Group I Contribution to the Intergovernmental Panel on Climate Change Fourth Assessment Report Climate Change 2007: The Physical Science Basis. Summary for Policymakers. 23pp

Jones HG (1992) Plants and microclimate: a quantitative approach to environmental plant physiology, 2nd edn. Cambridge University Press, p 452

Jurado E, Aguirre O, Flores J, Navar J, Jiménez J, Villalón H, Wester D (2000) Germination in Tamaulipan thornscrub of northeastern Mexico. J Arid Environ 46:413–424

Jurado E, García JF, Flores J, Estrada E (2006) Leguminous seedling establishment in Tamaulipan thornscrub of northeastern Mexico. For Ecol Manage 221:133–139

Lai L, Chen L, Jiang L, Zhou J, Zheng Y, Shimizu H (2016) Seed ger mination of seven deser t plants and implications for vegetation restoration. AoB Plants 8:plw031. https://doi.org/10.1093/aobpla/plw03

Meyer SE, Allen PS, Beckstead J (1997) Seed ger mination regulation in Bromus tectorum (Poaceae) and its ecological significance. Oikos 78:474–485

Nicotra AB, Atkin OK, Bonser SP, Davidson AM, Finnegan EJ, Mathesius U, Poot P, Purugganan MD, Richards CL, Valledares F, Van Kleunen M (2010) Plant phenotypc plasticity in a changing climate. Trends Plant Sci 15(12):684–692
Peñuelas J, Filella I (2001) Responses to a warming world. Science 294:793–794
Probert RJ (2000) The role of temperature in the regulation of seed dormancy and germination. In: Fenner M (ed) Seeds, the ecology of regeneration in plant communities, 2nd edn. CABI Publishing, Wallingford, pp 261–292
Reid N, Marroquin J, Beyer-Munzel P (1990) Utilization of shrubs and trees for browse, fuel wood and timber in the Tamaulipan thornscrub, northeastern Mexico. For Ecol Manag 36:61–79
Shimono Y, Kudo G (2005) Comparisons of germination traits of alpine plants between fellfilled and snowed habitats. Ecol Res 20:189–197
Solomon S, Quin D, Manning M, Chen Z, Marquis M, Averyt KB, Tignor M, Miller HL (2007) Climate change 2007: the physical science basis. In: Contribution of working group I to the fourth assessment report of the intergovernmental panel on climate change. Cambridge University Press
Vassilevska-Ivanova R, Tcekova Z (2002) Effect of temperature on seed germination and seedling growth of sunflower (Helianthus annuus L.). Comptes Rendus de l'Academie Bulg des Sci 55: 10–67
Villers-Ruiz L, Trejo-Vázquez I (1997) Assessment of the vulnerability of forest ecosystems to climate change in Mexico. Climate Res 9:87–93
Wang JH, Baskin CC, Chen W, Du GZ (2010) Variation in seed germination between populations of five sub-alpine woody species from eastern Qinghai-Tibet Plateau following dry storage at low temperatures. Ecolo Res 25(1):195–203

Presence and Importance of Mesquite *Prosopis laevigata* (Humb. & Bonpl. ex Willd.) M. C. Johnst in Northeastern Mexico

Horacio Villalón-Mendoza, Erica Esmeralda Hernández-Hernández, and Nelson Manzanares-Miranda

Abstract Mesquite *Prosopis laevigata* is an important tree in the arid southwestern United States and northern Mexico. It is a tree, native to Mexico, this is a very drought resistant. The interest in this species lies in the variety of uses and goods it provides, which range from fruit production, forage production, wood with excellent physical properties and notable calorific properties to be used as fuel and for medicinal use. The objective of this study is to know, through the systematization of information, on this species its ecological and socioeconomic impact in northeastern Mexico. It was possible to conclude that mesquite has a high ecological and socioeconomic impact in northeastern Mexico, comes to represent one of the species with more purposes for use by human beings and with a great impact on the ecological sustainability of the habitats where it interacts. It develops in the arid and semi-arid regions of Mexico and is very important for its wood, it is used as fuel, for the construction of fences, its pods as fodder and as food for man; produces resin that is used in the manufacture of glues and varnishes, while its flowers are important in the production of honey. Furthermore, mesquite is a biotic resource with wide geographic and ecological distribution in arid Mexican zones. This species has a very important ecological role, as it is an excellent soil fixer and therefore, erosion controller; it is a nitrogen fixer, which improves soil fertility

Keywords Prosopis · Native · Biology · Uses

H. Villalón-Mendoza (✉) · E. E. Hernández-Hernández
Facultad de Ciencias Forestales, Universidad Autónoma de Nuevo León, Linares, Nuevo León, Mexico

N. Manzanares-Miranda
Centro de Investigación en Producción Agropecuaria, Universidad Autónoma de Nuevo León, Linares, Nuevo León, Mexico

F. Garza-Ocañas (ed.), *Sustainable Management of Natural Resources*, Earth and Environmental Sciences Library, https://doi.org/10.1007/978-3-031-33394-1_8

1 Introduction

Mesquites are arboreal or shrub plants distributed in the arid, semi-arid and subtropical zones of different regions of the continental surface. They belong to the family Leguminosae, subfamily Mimosoideae and the genus *Prosopis*, which includes about 40 species (Frías Hernández et al. 2000).

Mesquite Prosopis laevigata is an important tree in the arid southwestern United States and northern Mexico. According to cronquist's classification system, it belongs to the Mimosaceae family of the order Fabales, class Magnoliopsida (dicotyledons).

It is a tree, native to Mexico, grows in semi-desert areas and with little rain. This is a very drought-resistant tree. It can measure up to 13 m. in height, its wood is very resistant and has thorny branches. The leaves of mesquite are bipinnate and narrow, measuring up to 7.5 cm. This tree has been used since the time of the Aztecs (Reinoso 2016).

Mexico has a wide extension of arid and semi-arid areas 56 and 23 million hectares, respectively, which, together, represent more than 40% of the total area of Mexican territory. We are talking about areas suitable for forest agricultural development based on the cultivation of mesquite. Undoubtedly, this is a resource that can be used to improve the living standards of the rural sector; it is currently established on more than 3.5 million hectares in northern Mexico.

The interest in these species lies in the variety of uses and goods they provide, which range from the production of fruits, whose nutritional properties make them suitable for human and animal consumption, wood of excellent physical properties to build furniture, crafts, construction materials, floors, and remarkable calorific properties to be used as fuel, its bark and leaves are used as healing remedies for different conditions, and in some species, the trunk and branches excrete a gum with similar characteristics like that of the Arabic gum (Frías Hernández et al. 2000).

In our country, despite being a plant resource of wide distribution and utility, it has not deserved the attention of the official and research sectors, which is reflected in a scarce existence of literature that addresses this topic, contrasting with what has been generated in countries such as Argentina, Brazil, Chile, India, the United States, etc., where aspects such as taxonomy have been addressed, ecology, productivity and others (Frías Hernández et al. 2000).

Terrestrial ecosystems present, by their very nature, a great ecological variability, so the mesquite is a species that by its geo-spatial coverage, usually by its size, its presence in arid and semi-arid areas, its morphological characteristics from its radical system to its foliage at the top of its crown, positively impacts the habitat, as it is through its leaf area index and vegetation cover through its large crown as an adult, its root depth that can reach more than 40 m deep depending on the soil characteristics of the site where it develops, fixation of N by its association with different species of bacteria fixing it, among others. The soil cover provided by this species in dry areas has direct effects mainly with the conservation of soil moisture, regulating temperature, which promotes biological activity in the soil and supports erosion control (Villalón-Mendoza 2014).

2 Objectives

The present study pretends to make available the knowledge, through the systematization of information on this species, the importance of the presence of mesquite (*Prosopis* ssp.) as a plant species, which, due to its ecological and socioeconomic impact in northeastern Mexico, it represents one of the species with more purposes of use by humans and of great impact on the ecological sustainability of the habitats where it interacts.

3 Methodology

To carry out this study, we proceeded to search for literature in secondary information sources to fully comply with the objective set out in this research.

4 Results

4.1 Species description

Common name: MESQUITE Family: FABACEAE

Scientific name: Prosopis laevigata (Humb. & Bonpl. ex Willd.) M.C. Johnst. Areas where it is present. Figure 1 shows that mesquite is well distributed in Mexico.

Mesquites are plants of arboreal or shrub stratum that grow in arid, semi-arid and subtropical areas of different regions of the Earth. These plants belong to the family Leguminosae, the subfamily Mimosoideae and the genus *Prosopis*. This legume is widely distributed in the dry regions of America, constituting frequency the only arboreal element of vegetation. Of the 42 species registered for the American continent, 9 can be found in Mexico, which can be found throughout the national territory except for the mountains and the rainy region of the southeast. It is estimated that its distribution is approximately 7 million hectares, concentrating mainly in the north and center of the country. The mesquite species recorded in northeastern Mexico are Prosopis laevigata (Humb & Bonpl. ex. Wild), M.C. Johnst. and *Prosopis glandulosa* Torr., with the *glandulosa* and *torreyana* (L.D. Benson) M.C. Johnst varieties. (Jiménez et al. 2005). Mesquite is considered a timber forest resource of economic importance because its wood is strong and durable, good for the manufacture of furniture, doors, windows, floors, handles for tools, posts for fences, as firewood and for the elaboration of charcoal, which has a national and

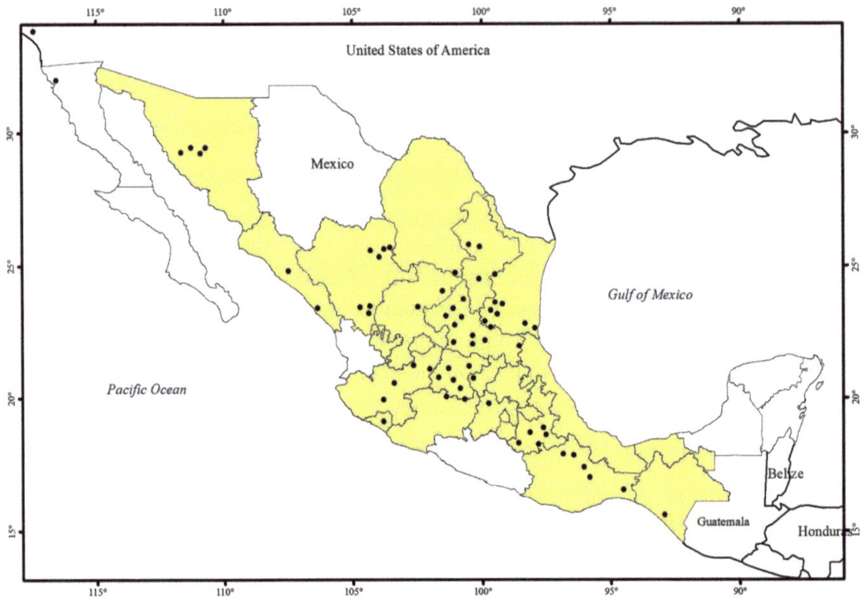

Fig. 1 Distribution of the main *populations of Prosopis* ssp. in Mexico

international demand. Mesquite pods and flowers are consumed by domestic livestock and wildlife; it is an important source of nectar and pollen to produce bee honey; in addition, mesquite gum can be a substitute for Arabic gum, of importance in the industry as an emulsifying agent, among other uses; the pods and seeds are used to make a flour for human consumption of excellent quality. From the ecological point of view, mosques are important in the structure and functioning of ecosystems, are habitat for wildlife, improve the aesthetics of the landscape and the functioning of hydrological basins (Mesa Sánchez 2009). In addition, it is considered a tree with ornamental potential of northeastern Mexico (Alanís-Flores 2005). Mesquites are an important part of the flora, even reaching a predominant character in certain regions; they have been linked to the life of the Mexican peasant since ancient times. Subjected to an irrational anthropogenic action since the Spanish conquest that has significantly deteriorated its natural populations, the genus *Prosopis continues* to be of great importance in the trophic chains of the ecosystems where it is distributed, serving as food and protection of wild fauna, as a soil stabilizer and protector of the hydrographic basins; however, like the rest of the species of wild flora typical of these regions, it has not been given the importance that it has as a development alternative. This species is a characteristic component of the Tamaulipas thorny scrub (Alanís-Flores 2007). There are 42 species reported in the state of Nuevo León and in the Sierra de San Carlos, Tamaulipas, which are used as sources of energy as fuel, of which only 8 are used to produce coal, within which there are 2 species of mesquite (*Prosopis* spp.), since its anhydrous bulk density is greater than 0.7 g/cm^3 (Villalón-Mendoza and Carrillo-Parra 2010).

4.2 Morphology and Anatomy of Mesquite

It is a tree, sometimes up to 12 m tall, although usually smaller; trunk up to 1 m in diameter, usually 30–60 cm; thick bark, brown blackish, somewhat fissured; cup wider than high; glabrous or hairy branches, armed with thorns stipulators 1–4 cm long; petiolate leaves with 1–3 pairs of pinnae, each with 10–20 pairs of sessile leaflets; flowers arranged in dense spikes 5–10 cm long; yellowish-white flowers; linear legume, somewhat falcate, 7–20 cm long by 8–15 mm wide, yellowish-brown, sometimes reddish, somewhat constricted between the seeds; they are oblong, compressed from 8 to 10 mm long, yellowish-white.

4.3 Mezquite Adaptive Strategies (Palacios et al. 2000)

The foliar *characteristics of Prosopis* that are considered adaptive strategies to thrive in the varied climatic and/or soil conditions present in the range of distribution of each species are the following:

At the morphological level

- Reduction of the leaf area, in the crop under controlled irrigation it could be observed that the size of the leaves is not a genetically fixed character but shows a strong positive correlation with water availability.
- Presence of extrafoliar nectaries: they are in the upper part of the petiole, on the primary or secondary rachis. Its activity increases during the midday hours (between 11 and 16 hs) when the temperature is maximum and the ambient humidity minimum. They are only functional in young leaves. It is considered that its function would be more related to the protection of young leaves against desiccation than to an anti-herbal action.

4.4 Floral Biology (Palacios et al. 2000)

The contributions for the knowledge of the ecology of pollination in the genus have been important, however, certain unclarified aspects persisted:

It was unknown whether the exert styles were really receptive, therefore, we were not sure that the flowers were really protogynous.

It was not known if the flowers in this phase were visited by insects.

The function of flowers with shorter style than the rest was unknown, a fact observed already by Burkart in 1937.

Pollinators were not known to be effective for mesquite-carob trees, as they receive about 20 different taxa when they are in full bloom.

The initial objective of the studies carried out has been to answer these unknowns.

So far 5 species have been studied over almost 10 years and for the moment it is possible to conclude:

From the beginning of the anthesis until the flowers fall, if there was no fertilization, the following floral phases are distinguished:

- Prefloral: closed corolla, protruding style, curved towards the apex of the inflorescence, non-receptive stigma, flower without nectar.
- Gradual opening of the corolla, anthers arranged parallel to the floral axis. Non-receptive stigma, flower without nectar.
- Corolla completely open. Anthers gradually placing perpendicular to the floral axis and beginning to open. Non-receptive stigma. Beginning of nectar secretion.
- Open anthers, receptive stigma (bubbling with hydrogen peroxide). Significant secretion of nectar. The flowers persist in this phase between 10 and 13–14 hs.
- Anthers have no pollen because of foraging by floral visitors. Stigma ceases to be receptive. There is no nectar, or it is very scarce. The wilting and falling of the flowers occur if there was no fertilization.

Therefore, stylar precocity does not imply the existence of protogyny.

4.5 Mesquite Reproductive System (Palacios et al. 2000)

It is necessary to point out an aspect of singular importance for any biodiversity management plan as well as with the collection, evaluation, improvement, and conservation of germplasm: the knowledge of the reproductive system. In these publications it is signed that in an experimental way the self-compatibility was verified in *P. laevigata* and *P. glandulosa* var. *torreyana*. The topic must be continued since it would not be uncommon to find differences between species or that in the boundary regions of the area of dispersion of these modifications in the reproductive system were detected. Establishing specific limits, ecological requirements, adaptive and reproductive strategies, is not only an academic issue, but also constitutes the fundamental tools for the knowledge of biodiversity in the genus and the use of its species in a rational and self-sustaining way. As can be understood, the task for the coming years is not easy, and today more than ever the good health of arid and semi-arid environments depends on the knowledge, understanding and management capacity that can acquire the largest predator and predator today, this is *Homo sapiens*.

The genus *Prosopis* in Mexico (Maldonado-Aguirre and de la Garza 2000a, 2000b).

Mesquites constitute an important part of the national flora, even reaching a predominant character in certain regions, they have been linked to the life of the Mexican peasant since ancient times.

The mesquite complex is firmly established on more than 3.5 million hectares of northern Mexico and includes the following native species:

P. laevigata var. *laevigata*. It is the typical mesquite of the Centro de México, it is the dominant species in San Luis Potosí, as well as in the center and south of Tamaulipas. It is distributed in isoyets from 300 to 900 mm. And at altitudes up to 2300 meters above sea level, mainly between 1800 and 1900, it is presented in the form of trees with heights of 6–7 m, as well as shrubs of 2–3 m, from its wood doors, furniture and carts are manufactured. Likewise, it is used for fuel and its fruits are used intensively as fodder for livestock, the fruit is also consumed as food by the inhabitants of the region.

P. glandulosa. It is one of the most widely distributed mesquite species in Mexico. Of this species are known the typical *glandulosa* variety that is distributed in Texas and in the Northeast of Mexico and the Torreyana variety that concentrates its distribution towards the Northwest portion on the Pacific coast, this variety is the most aggressive and mostly combated in the grassland areas. *P. juliflora* var. *juliflora*. It is distributed on plains, slopes and along streams and streams. *P. velutina*. It is distributed in northeastern Mexico, growing in alluvial wages, in the lower parts of beaches and along river and stream systems. *P. pubescens*. grows in northern Mexico in the states of Baja California, part of Chihuahua and Sonora. It grows in deep alluvial soils, along the soil banks of watercourses. The name of the species refers to the type of hairy fruit. *P. reptans*. It is a species of low size, it is distributed mainly in the Northeast region of Mexico, in alluvial soils, but with the presence of a layer of induration of calcium carbonate in its profile. The name of the species reptans means to crawl, which is observed in its growth form. *P. articulata*. It is distributed in a limited area, near Guaymas, Sonora and in the states of Tamaulipas and Veracruz, grows on rocky tables and on plains along the Baja California Peninsula. *P. tamaulipana*. Endemic species of Mexico that is distributed east of the Sierra Madre Oriental in the states of Tamaulipas and Veracruz, grows in clay soils and at low altitudinal levels. *P. palmeri*. Also endemic to Mexico whose distribution is limited to the Baja California Peninsula, it grows mainly in or near dry stream beds and desert beaches. The highest concentrations of mesquites are in the northern and central states of the country, among which Sonora, Chihuahua, Coahuila, Nuevo León, Tamaulipas, Durango, Zacatecas, San Luis Potosí, Guanajuato and Queretaro

Mesquite density varies from less than 10 trees per hectare in a crassicaule or grassland, to densities greater than 250 trees per hectare in an extra-desert mosque. In general, the average density is around 100 trees per ha. The microphyllus scrub with *Prosopis-Bouteloua* and *Opuntia*, in the state of Nuevo León, at altitudes of 130–340 mabove sea level and precipitation of 400–500 mm per year. Similarly, for Chihuahua, Nuevo León, Durango and for Coahuila, scrub is reported with Larrea in association with *Prosopis glandulosa* when the soil is deep, with plant heights of 1–2 m in the plains, while in the riverbeds it reaches 4 or 5 m in height. The root system of mesquite depends on the type of soil and the depth of moisture penetration. Young mesquites develop a strong pivoting root, subsequently generating the lateral root system. The characteristics of deep soil and climatic conditions of the distribution areas of the extra desert mosque have determined that a large part of them is occupied for agricultural purposes.

4.6 Properties and Characteristics of Mesquite

The mesocarp of the legume contains 13–36% sugars, while the seed has 55–59% protein. The fruit is sold to wholesalers in some municipalities of Tamaulipas. Mesquite produces up to 25 tons of fruit per hectare in certain places; it is estimated that in the same area up to 2 thousand kilograms of weight of cattle can be produced, while in the same area planted with corn only 563 kg would be obtained. Once transformed into flour, the fruit is used in the production of feed for cattle raised in stables. In the municipality of Matehuala, San Luis Potosí, the production of mesquite fruit has been estimated at seven thousand tons per year. In addition, its flowers are important producers of honey (Meraz et al. 1998). The wood of this tree is highly appreciated both for the manufacture of furniture and for the cooking of meats, to which it gives a pleasant flavor. From the resistant wood are made the sought-after parquet floors, as well as various handicrafts. Also, from the tree are obtained firewood and charcoal for domestic use, and poles for delimitation of plots (Meraz et al. 1998). The charcoal of this tree is distributed in the main cities of San Luis Potosí and Sinaloa. Its low production costs in Mexico have determined the increase of exports to the United States. Annual coal shipments to the northern neighbor went from 2000 to 20,000 tons from 1982 to 1992. An important example of the response to this market has been given by Sonora, where the production of mesquite coal was increased from 4 thousand tons in 1982 to more than 22 thousand in 1985, and its exports rose from 177 tons to more than 10 thousand annually in the same period (Meraz et al. 1998).

Since firewood and mesquite charcoal are highly prized products, the unquantified over-exploitation has caused serious deterioration, without measures being taken so far to regulate it. Through the practice of selective logging, many mosques have been destroyed; thus, in Nuevo León these plant formations are scarce and only some relicts can be found (Meraz et al. 1998).

From the ecological point of view, the different species of mesquite have a great importance because they serve as food and shelter for wildlife; they are also soil stabilizers and protectors of hydro-geographic basins.

These species are tolerant to salinity and there is evidence that they also can fix nitrogen. That is why FAO experts recommend them as producers of fodder and wood in the arid and semi-arid areas of the world. And it is that these are suitable plants for reforestation in the face of the desertification process, which is nothing more than the irreversible degradation – product of human activity – of large areas of land. In this way, various species of mesquite have been introduced in more than ten countries on all continents, except Europe (Meraz et al. 1998).

While in Mexico it is sought to optimize the use of its various qualities, in the United States research is oriented towards the development of better techniques for its combat, since they consider mesquite a weed in the places where intensive livestock is practiced (Meraz et al. 1998).

4.7 Uses of Mesquite

Mesquite is commonly considered an undesirable species, due to competition with pastures and decrease in forage production and increasing difficulty in managing livestock. However, the use of *Prosopis* by the inhabitants of northern Mexico dates to ancient times, having been very valuable as a resource of food, fuel, medicine, beverages, coal and construction and protection material (Frías Hernández et al. 2000).

In many regions of Mexico mesquite has been used in various ways due to its multiple qualities, its characteristics allow it to be used directly or transformed: mesquite is a magnificent soil improver, its leaves deposit an organic mulch of considerable importance, it fixes nitrogen to the soil, its roots control the movement of dunes, it provides fodder for domestic animals, serves as a habitat for wildlife, produces nectar for the production of bee honey, its fruits are consumed as food for humans, its wood is used as fuelwood, for rustic constructions, sleepers and poles (Frías Hernández et al. 2000).

An extremely important aspect within the problem of forestry activity is the extraction of wood for firewood and charcoal, according to an FAO study, 65% of the wood extracted in Mexico is used as firewood.

Currently, the inhabitants of almost all the rural communities of the arid and semi-arid region of the country consume firewood to solve their calorific requirements, having determined that at least 30% of the peasants make use of this resource (Frías Hernández et al. 2000).

Mesquite produces the best firewood that can be obtained in the semi-desert region. Its wood is heavy and dense, presenting a high calorific value. Its combustion is stable and constant, also imparting a pleasant flavor to food. Mesquite is therefore the preferred and essential energy resource among the inhabitants of the region. However, the constant use of this type of vegetation is what has further reduced their populations, so that their sources of use are being depleted rapidly, presenting a strong shortage of fuel with adequate calorific value (Frías Hernández et al. 2000).

Mesquite wood has certain characteristics that make it one of the best. Its hardness, color, texture, stability, and the beauty of its finish can be of primary importancc for its usc in furniture and crafts, manufacture of floors, as well as for many other uses (Frías Hernández et al. 2000).

The wood of *Prosopis* spp. It is somewhat harder than species such as oak (Quercus), cherry (*Prunus*) and walnut (*Juglans*), since it has a density greater than 0.7, but its real advantage is that its maximum volumetric change of 3 to 4% is among the lowest of any species in the world, which means that it has a lower tendency to expand or contract with changes in environmental humidity, than any other fine species.

Due to the characteristics of mesquite trees, in Mexico there are no commercial measures of wood, since its hardness, malformations, etc. They make it difficult to standardize and work, however, precisely those same characteristics make the articles made with that wood very appreciated among the lucky people who can acquire them (Frías Hernández et al. 2000).

Due to the characteristics and technical qualities of mesquite wood, it should have a price 5 to 10 times higher as wood than what is currently paid in the United States for the firewood of the species, therefore, the development and management of *Prosopis* should be seriously considered with the aim of producing wood for furniture or fine floors (Frías Hernández et al. 2000).

Mesquite charcoal has been an important product for domestic use, for many years and has a wide market acceptance, currently increasing its popularity for use in the kitchens of many homes, recreational areas and restaurants and other specialized establishments. A good quality coal has a calorific content of 31,000 Mj/kg (7.2 Kcal/kg). Hardwood coals have an ignition temperature of 300–450 °C. Mesquite charcoal is generally preferred over any other type of charcoal. It presents a relatively high amount of heat, which extends for a long period, being its combustion clean. The production of mesquite coal in Mexico was approximately 2000 tons per year until the year of 1980, when production rose to 10,000 tons by the year of 1985, the production of mesquite coal in Mexico increased to 35 thousand tons, of which about 20 thousand tons are exported annually to the United States (Frías Hernández et al. 2000).

The pods are used as fodder for livestock, these have a high nutritional value. The flour obtained from the mesquite pods can be mixed with corn flour, alfalfa, grass, flour, bran, ground alfalfa, peanut paste or flaxseed. Mesquite flour has a proportion of 20 to 60% of the total mixture that is used for fodder. It is also appreciated as a honey plant and obtaining gums for pharmaceutical uses. (Reinoso 2016). Pods were an important part of the diet in some states of the country. The seeds are used for a wide variety of meals and bread production. So, it is easy to integrate this excellent source of protein into our diet (Reinoso 2016).

According to Frías Hernández et al. (2000), along with mesquite foliage, pods are consumed by livestock, especially during droughts or forage shortages. In Mexico, thousands of tons of pods are collected annually from natural populations and sold as fodder or concentrate in rations for livestock. Pod production varies considerably between years, species, sites and even between trees of the same species.

It is estimated that the annual production of pod in the Altiplano varies from 4 to 50 kg/tree and from 200 to 2200 kg/ha, in densities of 25–445 trees/ha, however, it is considered that only 15% of the trees produce appreciable amounts of pod, so under certain techniques (genetic selection and planting of higher individuals, thinning and stand management) could greatly increase pod yields per hectare.

In Sonora the seeds were ground in a volcanic stone called "molcajete" and the flour was used to make atoles. Workshops have been held in which cooking is resumed with this ingredient for the preparation of sopes and gorditas (corn flour mixed with pig fat and some cooked food), empanadas, atole, among other foods.

To make mesquite flour, it is necessary to dry the pods in a mosquito net for 4–5 days. Then the pods are ground to obtain the flour. Mesquite flour is a source of protein, lysine, calcium, manganese, iron and zinc (Reinoso 2016).

It is interesting to know that, from legumes, some peoples in the dry regions of the country made intoxicating drinks. The fruit of mesquite is liked by all domestic ruminants and used for human consumption, as it has nutritional properties. The legume is consumed as fresh fruit; once dried, a farinaceous powder is obtained that can well be eaten like this, mesquite pinole, or used to make compact sweets (piloncillo), which are traded in regions of San Luis Potosí and Sinaloa.

On the other hand, they are also consumed by some small mammals and deer. The seeds, like many other legumes, are high in protein and carbohydrates.

4.8 **Prosopis** ***Is Widely Used For food and Wood Tools in the Region***

Natives of mosque areas in the past relied on mesquite honey and pods as a major component of their diets (Frías Hernández et al. 2000).

The abundant secretion of floral stigma allows it to be used in beekeeping, a mesquite tree can produce nectar for bees to make 1kg of good quality honey (Frías Hernández et al. 2000).

Mesquite pods contain large amounts of sugar and protein, so they are currently consumed as fresh fruit or boiled in their honey. Already dried, a sweet floury powder is obtained from them that can be consumed as pinole, cookies, bread, cake, in atole with milk, as a substitute for coffee, etc., or as compact sweets called piloncillos, which are the object of local trade (Frías Hernández et al. 2000).

It is worth mentioning that, supporting the importance, in Mexico, mesquite is widely used in reforestation areas with forestry and agroforestry vocation, in arid, semi-arid and subtropical ecosystems, since it provides great benefits to humanity. The process of benefiting their seeds involves a lot of work and loss of these in the mechanical process of extracting them from the fruit.

4.9 Lines of Action (Frías Hernández et al. 2000)

It is evident that the growing interest in mesquite and its products, as well as the destructive nature of its uses, together with the opening of land to cultivation, population growth, overgrazing, the presence of pests, diseases and fires, etc.

contributes strongly to the deterioration of the valuable resource, so it is necessary to establish action plans to generate and validate technologies for restoration, conservation and use of this important species to meet the demands of raw materials and other goods and services for the present and future times.

4.10 Forms of Exploitation (Frías Hernández et al. 2000)

The fruit of mesquite is consumed by all types of cattle: Mesocarp contains 13–36% sugars, while the seed has 55–59% protein. The quality of its wood is surpassed by few commercial species, firewood has excellent calorific qualities and the charcoal obtained is excellent (Leakey and Last 1980).

Prosopis species are tolerant to salinity and there is evidence that they also possess the ability to fix nitrogen. The cultivation and improvement of mesquite has been recommended for fodder and timber production (Burkart 1976, Leakey and Last 1980). While the FAO suggests its use for the reforestation of arid and semi-arid areas (Folliott and Thames 1983).

4.11 Natural Hybridization (Frías Hernández et al. 2000)

Analyzing the morphological variation in the mesquite of Tamaulipas and eastern Nuevo León, Frías Hernández et al. (2000) found individuals with intermediate characteristics, which suggests the occurrence of hybridization between *Prosopis glandulosa* var. *glandulosa* and *P. laevigata* and detected morphological signs of hybridization between various species of mesquite from Coahuila, Nuevo León, Sonora, and Baja California Sur.

4.12 Main Forms of Exploitation (Frías Hernández et al. 2000)

1. In human food
   The pod is consumed as fresh fruit: already dried, a farinaceous powder is obtained from it that can be eaten like this (mesquite pinole) or used to make compacted sweets. (e.g. mesquite piloncillo's) that are the object of local trade, with the fermentation of the sugars of the fruit an alcoholic beverage is made, the amber gum secreted by the trunk of the tree, is a treat for children.
2. In traditional medicine
   Using foliage, rubber and bark, its use as an antidysentery, antidiarrheal, emollient, and anti-obstetric antiseptic was recorded.

3. Forage use

   The pod is very desired by all kinds of cattle. The rural population collects it dry for their animals and can store it for periods of drought. In 1982 the commercial collection of mesquite fruit exceeded three thousand tons in the municipality of Matehuala, once converted into flour, it is used to make balanced feed for stabled cattle.
4. Forest exploitation

   Firewood and mesquite charcoal are highly prized products; however, they have been the cause of the serious deterioration of mesquites. The firewood is collected for self-consumption and as an object of trade, the charcoal is distributed to the main cities of the region.

   Furniture of beautiful finish and aesthetic parquet floors are made with the resistant mesquite wood, being very well valued, also various crafts are produced.

   Mesquites are important honey species. There are some parks in the region shaded mainly by mesquite trees.

   The forms of mesquite exploitation recorded in this research are fundamentally similar to those practiced in the rest of the world where *Prosopis* species grow (Burkart 1952; Leakey and Last 1980).
5. Other types of use (Frías Hernández et al. 2000)

   Considering the economic, ecological, and social aspects, in the use of mesquite priority should be given to the collection and treatment of the fruit as fodder, as well as to the elaboration of furniture, parquet floors and handicrafts. It would be advisable to implement effective reforestation programs with mesquite, in addition to experimentally trying to cultivate and improve mesquite as a fodder tree. The above on the basis that:

   - Tree mesquites induce substantial changes in the characteristics of herbaceous vegetation. The plant under its canopy differs significantly from that found in open areas and is dominated by perennial or annual, caespitous and short grasses. The change in vegetation is abrupt and has as its limit the vertical projection of the canopy.
   - The herbaceous species that tend to settle under the mesquite canopy are not always the same, observing a certain level of exclusivity which indicates the presence of competitive hierarchies. However, the presence of one or another species seems to be the result of random events that influence their ability to disperse.
   - Shrub mesquites that form conglomerates and have very closed canopies, have a very scarce herbaceous vegetation (specially grasses). The latter can be explained by low levels of radiation.

5 Conclusions

Mesquite is developed in the arid and semi-arid regions of Mexico, including northern Sinaloa and is very important since its wood is used as fuel, for the construction of fences, its pods as fodder and as food for man; it produces resin that has use in the manufacture of glues and varnishes, while its flowers are important in the production of honey.

Mesquite is a biotic resource with wide geographical and ecological distribution in arid Mexican areas. This species has a very important ecological role since it is an excellent soil fixator and therefore, an erosion controller; it is a nitrogen fixer, which improves soil fertility. On the other hand, under certain conditions they are a source of fodder for domestic livestock and wildlife.

The results obtained in terms of protein content, amino acid profile, relative protein value, in vitro digestibility of the protein and energy value indicate that the whole pod of *Prosopis laevigata* has nutritional qualities like other legumes, so it suggests a great potential for use and use in food products for humans.

The pod is attacked by the Bruquids, *Prosopis laevigata* by *Algarobius* spp. The damage of these beetles to the seed ranges from 44% or more in *Prosopis laevigata*. The damage of these insects to the mesquite seed should be considered in the reforestation programs with this plant, as an alternative in the biological control of a thorny herb called "gatuño", since every day this plant invades more cattle areas, with respect to the huizache the number of species found in its seeds indicate the diversity of entomofauna that they host. These plants, which should be considered as reservoirs of diversity so damaged lately. The percentage of germination is favored when the mesquite seed is scarified through the digestive tract of goats and cattle compared to horses, donkeys and the control.

Carbon stocks in the soil where mesquites develop up to one meter deep range from 40 t/ha in arid zones to 80–100 t/ha in subtropical and tropical zones. The arid zones in which mesquites grow cover 40% of the global land area, storing 5% of the total carbon stored in the world's forests.

The soils under the mesquite canopy are superior in abundance of bacteria, fungi, arbuscular mycorrhizal fungi and rhizobial bacteria, they also showed greater activity (measured in$_{CO_2}$ release), the above among other aspects, evidence the importance of mesquites in the ecological balance of semi-arid ecosystems.

References

Alanís-Flores G (2005) Plantas Nativas del Noreste de México con Potencial Ornamental. Tópicos Selectos de Botánica 2. Etnobotánica, sistemática, fisiología y plantas en ambientes urbanos. Dpto. de Botánica. Facultad de Ciencias Biológicas, U.A.N.L. Monterrey, Nuevo León, México. pp 161–173

Alanís-Flores G (2007) Perspectivas Ecológicas y Económicas en el Manejo de Recursos Vegetales del Noreste de México. Tópicos Selectos de Botánica 3. Etnobotánica, sistemática, fisiología y plantas en ambientes urbanos. Dpto. de Botánica. Facultad de Ciencias Biológicas, U.A.N.L. Monterrey, Nuevo León, México. p 37

Burkart A (1952) Las leguminosas argentinas, silvestres y cultivadas, 2da. edn. Acme, Buenos Aires

Burkart A (1976) A monograph of de genus *Prosopis* (Leguninosae, Moimosidae). J Arnold Arbor 57(3):217–249;57(4):450-525.

Folliott P, Thames JL (1983) Manual sobre taxonomía de *Prosopis* en México, Perú y Chile, FAO, pp 1–35

Frías Hernández JT, Olalde Portugal V, Vernon Carter J (2000). El Mezquite. Árbol de Usos Múltiples. Estado Actual del Conocimiento en México. Universidad de Guanajuato, México. 247p

Jiménez VA, Torres CE, Foroughbakhch R, Alvarado VM, Treviño-Garza, EJ (2005) El Mezquite (*Prosopis glandulosa* Torr.) en el Noreste de México. Tópicos selectosde Botánica 2. Etnobotánica, sistemática, fisiología y plantas en ambientes urbanos. Dpto. de Botánica. Facultad de Ciencias Biológicas, U.A.N.L. Monterrey, Nuevo León, México. p 61

Leakey R, Last T (1980) Biology and potential of *Prosopis* species in arid environments, with particular reference to *P. cineraria*. J Arid Environ 3:9–24

Maldonado-Aguirre LJ, de la Garza P (2000b) El Mezquite en México: Rasgos de Importancia productiva y Necesidades de desarrollo. En: Frías H.J.T., Olalde P., V. y J. Vernon C. El Mezquite. Árbol de Usos Múltiples. Estado Actual del Conocimiento en México. Universidad de Guanajuato, México. pp 37–50

Maldonado-Aguirre LJ, de la Garza, PFE (2000a) El mezquite en México: Rasgos de importancia productiva y necesidades de desarrollo. En: Frías Hernández, Maldonado- Aguirre, L.J. y de la Garza de la P., F.E. 2000. El Mezquite en México: Rasgos de Importancia Productiva y Necesidades de Desarrollo. El mezquite, árbol de usos múltiples. Estado actual del conocimiento en México. Universidad de Guanajuato. México. pp 37–50

Meraz V, Orozco V, Lechuga C, Cruz S., Vernon C. (1998). El mezquite, árbol de gran utilidad. Ciencias 51, julio-septiembre, 20–21

Mesa Sánchez R (2009) Guía para la colecta y beneficio de semilla de mezquite. Centro de Investigación Regional Noroeste. Campo Experimental Todos Santos, INIFAP. Folleto para Productores Núm. 2 junio de 2009. 36p

Palacios RA, Hoc PS, Burghardt AD, Viela AE (2000) *Prosopis* L.: Biodiversidady Clasificación, Estrategias Adaptativas, Reproducción e Importancia Económica. En: Frías H.J.T., Olalde P., V. y J. Vernon C. El Mezquite. Árbol de Usos Múltiples. Estado Actual del Conocimiento en México. Universidad de Guanajuato, México. pp 13–35.

Reinoso, V. (2016). Mezquite y Huizache, árboles mexicanos que fertilizan nuestro suelo. Octubre 10, 2018, de Asociación de los consumidores orgánicos Sitio web: https://consumidoresorganicos.org/2016/09/20/mezquite-huizache-arboles-mexicanos-fertilizan-nuestro-suelo/

Villalón-Mendoza H (2014) Capítulo 1. Técnica Bidimensional de Puntos de Contacto: un Método de Medición de la Cobertura Vegetal, Cobertura Edáfica e Índice del Área Foliar para Diferentes Ecosistemas Terrestres. Garza Ocañas F. (Editor) Técnicas en el Manejo Sustentable de los Recursos Naturales. Cuerpo Académico "Manejo de Recursos Naturales y Sustentabilidad". Edit. Universidad Autónoma de Nuevo León. Impreso en Monterrey, N.L., México. 2014. pp 14–25

Villalón-Mendoza H, Carrillo-Parra A (2010) Producción de Leña y Carbón. De la Lechuguilla a las Biopelículas Vegetales. Las Plantas Útiles de Nuevo León. Dpto. de Botánica. Facultad de Ciencias Biológicas, U.A.N.L. Monterrey, Nuevo León, México. p 277

Edible Macromycetes of Chihuahua. Diversity and Nutritional Properties

María Haydeé Solano Zavala, Ivette Molinar Monsivais, David Reyes Ruvalcaba, Miroslava Quiñónez Martínez, Irma Delia Enriquez Anchondo, and Fortunato Garza-Ocañas

Abstract Chihuahua (México) has a forest area of 30%, with plant communities dominated by pine and oak species, which are economically exploited for their timber quality. Almost 500 species of macroscopic fungi grow in these areas, which are ecologically important for maintain a healthy ecosystem due to its symbiotic, parasitic, or saprophytic relationships with other organism and abiotic components. These are also an important non-timber resource for local communities because some macromycetes are considered edible for its nutritious value or alleged medicinal properties. Currently, studies on nutritional analysis, antioxidant activity and antimicrobial capacity has generated useful and beneficial knowledge to promote its widely consumption and potential benefits on health and diet. The present study aims to identify the species of macromycetes fungi from temperate forests in the state of Chihuahua and relate it to relevant information regarding the nutritional and functional properties that allow its consumption and use.

Keywords Fungus · Edible · Health · Chihuahua · Forest

1 Introduction

Chihuahua is the largest state in Mexico, 70% of its territory is covered by forest area (Herrera 2022). The vegetation is dominated by coniferous (*Pinus* spp.) and oak (*Quercusk* spp.) commuties, which have and important economical, touristic, and cultural value for local resident (Quiñonez-Martínez et al. 2010).

Among these forests a diversity of fungi forms grows, whose ecological importance is variable. Saprobic forms are found that degrade organic matter, recycling

---

M. H. S. Zavala · I. M. Monsivais · D. R. Ruvalcaba · M. Q. Martínez (✉) · I. D. E. Anchondo
Instituto de Ciencias Biomédicas, Universidad Autónoma de Ciudad Juárez, Ciudad Juárez, Chihuahua, Mexico

F. Garza-Ocañas
Universidad Autónoma de Nuevo León, Campus Linares, Facultad de Ciencias Forestales, Linares, Nuevo León, Mexico

F. Garza-Ocañas (ed.), *Sustainable Management of Natural Resources*, Earth and Environmental Sciences Library, https://doi.org/10.1007/978-3-031-33394-1_9

nutrients from the forest floor. The parasitic forms live by infecting living trees of many pines, oaks, and other fungi (Aguirre-Acosta et al. 2014). Finally, ectomycorrhizal fungi stand out for their ecological contribution to the plant communities where they grow, since they form a symbiotic association with the roots of the tree and shrub vegetation of the forest, generating mutual benefits for both species (eg exchange of nutrients, water uptake, protection against parasites) (Rentería-Chávez et al. 2017).

Macromycetes have been consumed since ancient times due to their flavor, texture, nutritional content, medicinal and antimicrobial properties (Gysling-Caselli et al. 2005; López et al. 2017; Wasser 2002) nowadays are considered one of the most important non-timber resources provided by forest communities. Studies like Martínez-Carrera et al. (2004, 2012), Naranjo et al. (2013), Gómez-Flores et al. (2019) and Pontón (2008) have shown that edible mushrooms are nutritious foods and can contribute to supply macro and micronutrients in the diet. Although in Chihuahua around 450 species of macromycetes mushrooms have been registered, only a few species are used for residents as food (e.g. A*garicus silvestris* and *Amanita caesarea*) or by its medicinal properties (e.g. *Lycoperdon* sp.) (Sánchez and Mata 2012; Quiñónez-Martínez et al. 2014). The objective of this study is to present a list of fungal species considered edible, as well as their ecological function in diverse sampled plant communities distributed in different municipalities of Chihuahua, Mexico.

2 Methodology

2.1 Study Zone

Field trips were carried out from 2005 to 2019 through the forests of different localities of Chihuahua, state, in the municipalities of Bocoyna, Guachochi, Madera and Urique. The altitude range was between 1600 to 2400 ma.s.l. The rainy season occurs between June and October. The plant communities are constituted by pine forest, pine-oak forest, oak-pine forest, and mixed forest, with a variete of tree species such as *Pinus arizonica*, *P. engelmannii*, P. *durangensis*, P. *leiophylla*, *Quercus arizonica*, *Q. chihuahuensis*, *Q. jonesii*, Q. *mcvaughii*, *Q. crassifolia*, Q. *depressipes*, *Q. durifolia*, *Q. hypoleucoides*, *Arctostaphylus pungens* and *Arbutus xalapensis* (Quiñónez et al. 2008; Quiñónez-Martínez et al. 2008, 2014). Soils were shallow belonging to the Ferozems and Lithosols groups, characterized by a thin horizon layer and little organic matter (humus). Economic activities in the region include agriculture, forestry and tourism (INEGI 2013).

2.2 Field and Laboratory Studies

In each field trip, all possible macromycete fungi were collected through directed sampling. Subsequently, they were individually placed in a paper bag, and transported to the ICB-UACJ Biodiversity Laboratory. Macro and microscopic characterization were carried out. The works of Arora (1991); Becker (1992); Bessette et al. (2000); Guzmán (2008); Pompa-González et al. (2011), Quiñonez-Martínez et al. (2010) and Quiñónez-Martínez et al. (2014) were used to determine the species, ecological role, and edibility. Finally, a list with the edible species classified by ecological role is presented, also a literature review was conducted to summarize the nutritional profile of some representative species.

3 Results and Discussion

3.1 Species Richness

A richness 84 species of mushrooms considered edible was recorded, the information is presented in Table 1. The species that appeared most frequently in most of the sites are: *Amanita rubescens*, *Lactarius indigo*, *Cantharellus cibarius*, *Higrophorus russula*, *Laccaria laccata*, *Laccaria bicolor*, *Boletus pinophilus*, *Suillus granulatus, Helvella crispa*, *Hipomyces lactifluorum* and *Lycoperdum echinatum*. The greatest diversity of species belongs to fungi considered ectomycorrhizal ($n = 47$); the genera *Amanita*, *Boletus*, *Cantharellus*, *Laccaria*, *Lactarius*, *Russula* and *Suillus* stand out for their richness. 22 species of edible fungi considered saprobes were recorded. Lastly, five parasitic species were found, being *H. lactiflorum* the most common; *Ustilago maydis* was recorded in maize cultivars in the municipality of Bocoyna. Some of the species found and of commercial value are: *Agaricus campestris*, *Helvella crispa* and *Ganoderma applanatum*.

In total, there are more than 350 species of edible wild mushrooms in the country (Ruan-Soto and Mariaca 2012), our study reports 24% of the national richness. Similarly, it was possible to increase the number of reported and identified edible mushrooms for the state of Chihuahua by 63%, according to data from the study conducted by Quiñonez-Martínez et al. (2010), where up to 50 fungi with food potential discovered up to that date were mentioned. It has been previously mentioned that, despite the wealth of edible mushrooms in the state, local people prefer only a small group of species, including A. *rubescens*, *A. campestris*, *U. maydis*, *H. lactifluorum* and *A. caesarea* (Quiñónez-Martínez et al. 2014). In addition to their edible value, it has been shown that some of the species present in Table 1 have medicinal and antimicrobial potential, including *A. hygrometricus*, *L. laccata*, *G. lucidum*, *G. aplanatum* and *Lycoperdon* spp. (Martínez-Escobedo et al. 2021).

On the other hand, the species of the genus *Suillus*, *Boletus*, *Laccaria*, *Lactarius*, *Russula* and *Astraeus* have also been recognized for their high efficiency mycorrhizing seedlings and trees with timber forestry interest, promoting their development, settlement, and survival in wild environments (Martín-Pinto et al. 2006; Curguz et al. 2010; Kaewgrajang et al. 2013; Wang et al. 2015) (Fig. 1).

Table 1 Species reported with potential edible use of Chihuahua forests

Ectomicorrhizal	*Saprotrophic*	*Parasites*
Amanita caesarea	*Agaricus augustus*	*Auricularia auricula*
A. fulva	*A. campestris*	*A. polytricha*
A.hemibapha	*A. arvensis*	*Ganoderma aplanatum,*
A.vaginata	*A. bisporus*	*G. lucidumn*
A. rubescens	*A. silvicola*	*Armillaria mellea*
A. tuza	*Aleuria aurantia*	*A. tabescens*
Boletus appendiculatus,	*Calvatia cyathiformis*	*Hericium erinaceus*
B. bicolor,	*Clavariadelphus truncatus*	*Hypomyces lactifluorum*
B. chrysenteron	*Clitocybe gibba*	*Phellinus robustus*
B. edulis	*Coprinus comatus*	*Pleurotus ostreatus*
B. luridus	*Hygrocybe miniatus*	*Pleurotus pulmonaris*
B. mirabilis	*Lycoperdon echinatum*	*Sparapsis crispa*
B. palidus	*Lycoperdon perlatum*	*Schizophyllum commune*
B. pinophilus	*Lycoperdon pyriforme*	*Tremella mesenterica*
B. variipes	*Lyophyllum decastes*	*Ustilago maydis*
B. reticulatus	*Macrolepiota procera*	
B. rubellus	*Macropodia macropus*	
Boletellus russellii	*Melanoleuca sp.*	
Leccinum auriantacum	*Morchella esculenta*	
L. manzanitae	*Peziza hemisphaerica*	
Strobilomyces confusus	*Neolentinus ponderosus*	
Suillus americanus	*Pleurotus ostreatus*	
S. brevipes		
S. granulatus		
S. lakei		
S. pictus		
S. pseudobrevipes		
Cantharellus cibarius		
Cantharellus cinnabarinus		
Craterellus fallax, C.cornucopioides		
Hydnum repandum		
Gomphus clavatus, G. floccosus		
Helvella crispa		
Helvella lacunosa		
Hygrophorus russula		
Laccaria amethystina		
Laccaria bicolor		
Laccaria laccata		
Lactarius deliciosus		
Lactarius indigo		
Lactarius volemus		
Russula brevipes		

(continued)

Table 1 (continued)

Ectomicorrhizal	*Saprotrophic*	*Parasites*
Russula cianoxantha		
Russula delica		
Chroogomphus vinicolor		
Melanogaster variegatus		
Tricholoma magnivelare		

Fig. 1 Some edible species from the temperate forests of Chihuahua

3.2 Nutritional Composition

Edible mushrooms have a wide variety of nutritional properties, Table 2 shows a list of macromycetes species of wild edible mushrooms found in the study zone with their proximate analysis (%) reported form other authors. These indicate moisture (water content in the fungus), protein, carbohydrates, lipids (fats), crude fibers and ashes (total mineral content in the sample). In general, the results highlight an important source of protein, carbohydrates, as well as being low in fat, in addition to providing minerals and fiber to the diet. From the table, only *A. ceasarea*, *A. rubescens*, *Boletus edulis*, *B. pinophillus* and *H. lactifluorum* have been reported as consumed on the study regions, leaving a wide spectrum of fungal species for potential consumption, which may represent a supplemental source of nutrients.

Table 3 shows 30 species of edible wild mushrooms, some nutritional values (% dry weight) and the country in which they are mainly consumed. Following the example of other countries, residents of different communities in Chihuahua can start implementing into their diet new species of wild edible mushrooms.

Mushroom nutrition is a process that involves extracellular digestion, transport of the nutrients it absorbs, and metabolism. Fungi absorb the soluble organic nutrients they need as an energy source through their chitin wall and membrane (heterotrophs) by secreting enzymes into the surrounding environment and then absorb the nutrients. Fungi, apart from being heterotrophs, are also considered chemoorganotrophs because they need organic compounds as an energy source. Other fungi are oligotrophs as they grow in environments where nutrients are scarce, absorbing volatile organic compounds from the atmosphere. Due to the large number and variety of enzymes that fungi can synthesize, they degrade a variety of simple and

Table 2 Some studies on proximal analysis of edible mushroom species (% of fresh weight)

Species	Moisture	Ashes	Fibers	Proteins	Carbohydrates	Fats
Aricularia auricula[1]	92.43	3.36	**	9.25	47.32	0.41
Amanita caesarea[3]	93.4	12.0	20.2	14.7	48.6	4.9
Amanita rubescens[5]	92.9	6.78	19.9	17.4	47.6	8.3
Boletus edulis[3]	85.10	7.0	18.8	14.8	54.9	1.8
Boletus edulis[4]	92.26	4.64	-	13.3	40.1	1.7
Boletus frostii[5]	94.5	3.23	30.2	15.8	47.0	3.68
Boletus pinophillus[4]	92.47	5.49	-	12.6	36.3	1.72
Cantharellus cibarius[4]	93.89	0.88	-	0.63	4.18	0.43
Laccaria laccata[4]	92.19	17.8	-	7.34	34.3	1.32
Lactarius deliciosus[2]	92.70	9.60	10.60	7.3	43.8	4.9
Lactarius indigo[5]	95.1	4.93	18.7	13.42	58.6	4.25
Hypomyces lactiflorum[4]	91.55	0.87	**	0.90	6.45	0.24
Pleurotus ostreatus[3]	82.90	5.80	24.20	9.2	55.9	0.43
Ramaria flava[3]	93.2	6.8	26.4	15.5	10.42	3.25

1. Ríos-Hurtado et al. (2005); 2. Barros et al. (2009), 3. Naranjo et al. (2013), 4. Díaz-Sierra (2009); 5. León-Guzmán et al. (1997)

Table 3 Nutrient composition of some edible wild mushrooms in 100 g of dry weight

Species	Country	Proteins	Carbohydrates	Fats
Amanita caesarea	Francia	15	D	14
Amanita loosii	República Democrática del Congo	20	D	D
Amanita rubescens	Mexico	18	D	D
Botelus edulis	Turquía	38	47	9
Botelus edulis	Finlandia	23	Ns	2
Botelus erythropus	Jordania	15	57	1
Botelus frostii	México	16	Ns	Ns
Botelus loyo	Chile	22	50	1
Cantharellus cibarius	Turquía	21	62	5
Cantharellus cibarius	República Democrática del Congo	15	64	5
Lactarius phlebophyllum	República Unida de Tanzania	30	51	9
Lactarius delicious	Francia	23	Ns	7
Lactarius delicious	Chile	27	28	7
Lactarius indigo	México	13	Ns	Ns
Lactarius torminosus	Finlandia	21	Ns	2
Lactarius peperatus	Turquía	27	65	2
Ramaria flava	México	14	Ns	D
Ramaria flava	Finlandia	24	Ns	2
Russula cyanoxantha	Francia	17	Ns	8
Rusuula delica	India	17	Ns	Ns
Russula sp.	República Democrática del Congo	29	55	6
Suillus luteus	Chile	20	57	4
Suillus granulatus	Chile	14	70	2
Terfezia claveryi	Iraq	8	17	Ns
Terminomyces microcarpus 1	República Unida de Tanzania	49	29	10
Terminomyces microcarpus 2	República Unida de Tanzania	35	37	6
Terminomyces microcarpus	República Unida de Tanzania	33	38	5
Tricholoma populinum	Cánada	13	70	9
Tricholoma saponaceum	Francia	5	D	7
Tirmania nivea	Iraq	14	21	D

Ns: Not noted; D: unknown Source: Boa (2005)

complex compounds where they take advantage of them as an energy source (Walker and White 2017).

Nutritional requirements act dynamically according to their nutritional environment and can make physiological and morphological changes depending on the availability of certain nutrients (Walker and White 2017). Physiologically, when nitrogen or carbon sources are scarce, some fungi degrade their cellular components (autophagy) using enzymes or autophagosomes (Alexopoulos et al. 1996). The physical requirements for growth are temperature, humidity, light, pH and aeration (Gaitán-Hernández et al. 2006). These can lead to changes on the proximate analysis between regions, so specific studies are necessary for the potential edible species registered on this study.

Micronutrients are those that require in concentrations of 10-6 or less than these. These elements are enzymatic activators and participate in vitamin synthesis, sporulation, and growth. Among the microelements that the fungus needs are iron, zinc, copper, manganese, calcium, cobalt, molybdenum, and gallium (Walker and White 2017).

As for vitamins, they are considered as growth factors because they are organic compounds required in minimal quantities, sometimes these can be auto-synthesize with only glucose supply. The vitamins that are most frequently supplemented for fungal growth are: thiamine-B1, pyridoxine-B6, biotin and nicotinic acid (Giler et al. 2017; Walker and White 2017).

Cellulose is the most abundant polymer on earth. It consists of long unbranched chains of between 8000 and 12000 units of D-glucosapyranose with β-1-4 -glycosidic bonds. Fungi use cellulose as a carbon source. Parasitic fungi such as *Pleurotus ostreatus* have an enzyme complex to completely degrade complex carbohydrates into glucose. Endoglucanase (endo-β-1-4-glucanase) breaks the chain into random regions, celliohydrolase acts at the terminal ends of the chain by converting it into units of the disaccharide celobiose, and β-glucosidase or celobiose, hydrolyzes the celobiose to glucose. Finally, as a monomer it is absorbed by the fungus (Seidl-Seiboth et al. 2014). Lignin is a polymer considered to be the most abundant biopolymer. It is composed of oxyphenylpropanoid units connected by several different junctions between C–C and C–O. It is formed by the random binding of three monomeric precursors: coniferyl, sinapil (syringyl propanol) and ρ-coumaril (ρ-hydroxyphenylpropanol). It lacks glycosidic bonds, so its degradation is not by hydrolysis but by oxidation. *P. ostreatus* also have a system of peroxidase enzymes and laccases that completely degrade lignin. Laccases are multi-copper proteins with phenoloxidase activity that catalyze the oxidation of phenols and aromatic amines (Seidl-Seiboth et al. 2014). Regarding chitin, it is a polysaccharide composed of N-acetyl-glucosamine units with β-1-4 bonds. This is the main compound of the fungal wall (Seidl-Seiboth et al. 2014). There are currently many publications that mention the antioxidant capacity of edible mushrooms (Agrahar-Murugkar and Subbulakshm 2005; Elmastas et al. 2007; Álvarez-Parrilla et al. 2007; Barros et al. 2007, 2009; Gençcelep et al. 2009), highlighting species that commonly develop in forests, such as: *Cantharellus cibarius, Russula cianoxatha, Boletus edulis, Tricholomopsis rutilans* and *Suillus granulatus*, some of them with gastronomic relevance.

Some current studies have recognized that fungi contain an important source of biological active compounds of medicinal value (Gezer et al. 2006) and a variety of secondary metabolites including phenolic compounds (polyphenols, tannins, flavonoids, phenolic acids, stilbenes, etc.), as excellent synergistic sources that provide health benefits associated with the reduction of risks in degenerative and chronic diseases (Gezer et al. 2006). Some species with these compounds are found in: *Lycoperdon perlatum, Macrolepiota procera, Agaricus silvicola, Cantharellus cibarius, Ramaria flava, R. botrytis*, among others (Barros et al. 2009). Many of these fungi have been a source of bioactive compounds with antifungal, aintihypertensive, anti-inflammatory, antioxidant, and antiviral properties (Pérez-Moreno et al. 2010). An example is the study of bioactive compounds in *Auricularia auricula* carried out by Ríos-Hurtado et al. (2005); the authors reported the presence of steroids, triterpenoids and quinones, suggesting it as a therapeutic applicability. Likewise, their values of micronutrients and vitamins stand out, compared to some vegetables, such as peas, beans and others.

4 Conclusions

Edible wild mushrooms are an extensive group of organisms that possess a wide variety of nutritional properties, rich in amino acids, vitamins, essential minerals. They are also generators of secondary metabolites such as lecithins, phenolic compounds, polyphenolics, polysaccharides, among others, and contain antioxidant and lignolytic enzymes. Likewise, these are an important part of the economy of rural forest-related communities, as well as vital part of their diet, especially in rainy seasons (from June to September). The study also presents some edible macromycetes with attributed medicinal properties such as: antitumor, anticancer, antioxidant among others. It is important to continue with explorations in the forests of Chihuahua to update the richness of fungi with potential use (ecological, food, medicinal) for the inhabitants of the region and for the conservation of their forests.

References

Agrahar-Murugkar D, Subbulakshm G (2005) Nutritional value of edible wild mushrooms collected from the Khasi hills of Meghalaya. Food Chem 89:599–603

Aguirre-Acosta E, Ulloa M, Aguilar S, Cifuentes J, Valenzuela R (2014) Biodiversidad de hongos en México. Rev Mex Biodivers 85:76–81

Alexopoulos CJ, Mims CW, Blackwell M (1996) Characteristics of fungi. In: Introductory micology, 4th edn. Wiley, Milan, MI, pp 26–56

Álvarez-Parrilla E, De la Rosa LA, Martínez R, González GA (2007) Total phenols and antioxidant activity of commercial and wild mushrooms from Chihuahua, México. Soc Mex Nutr Tecnol Alim 5(5):329–334

Arora D (1991) All that the rain promises and more. California, Ten Speed Press: California, USA. 259p

Barros L, Baptista P, Correira DN, Casal S, Oliveira B, Ferreira ICFR (2007) Fatty acid and sugar compositions and nutritional value of five wild edible mushrooms from Northeast Portugal. Food Chem 105:140–145

Barros L, Dueñas M, Ferreira ICFR, Baptista P, Santos-Buelga C (2009) Phenolic acids determination by HPLC-DAD-ESI/MS in sixteen different Portuguese wild mushrooms species. Food Chem Toxicol 47:1076–1079

Becker G (1992) Setas. Susaeta Ediciones: Madrid. 223p

Bessette AE, Roody WC, Bessette AR (2000) North American boletes. University Press: Syracuse. 396p

Boa E (2005) Los hongos silvestres comestibles. Perspectiva global de su uso e importancia para la población. Roma, Italia. FAO

Curguz VG, Raicevic V, Tosic MT, Veselinovic M, Jovanovic LJ (2010) Same physiological characteristics of the three ectomycorrhizal fungi from *Suillus* genus. Min Biotecnol 22

Díaz-Sierra M (2009) Determinación de la composición proximal, fenoles totales y capacidad antioxidante de hongos silvestres comestibles de Chihuahua, México. Tesis de Licenciatura en Química. Instituto de Ciencias Biomédicas, Universidad Autónoma de Ciudad Juárez

Elmastas M, Isildak O, Turkekul I, Temur N (2007) Determination of Antioxidant Activity and Compounds in Wild Edible Mushrooms. J Food Compos Anal 20:337–345. https://doi.org/10.1016/j.jfca.2006.07.003

Gaitán-Hernández R, Salmones D, Pérez Merlo R, Mata G (2006) Manual práctico del cultivo de setas: aislamiento, siembra y producción, 1era. ed., 2a. reimp. Instituto de Ecología, A.C. Xalapa, Ver., México, 56pp

Gençcelep H, Uzun Y, Tunçtürk Y, Demirel K (2009) Determination of mineral contents of wild-grown edible mushrooms. Food Chem 113:1033–1036

Gezer K, Duru ME, Kivrak I, Turkoglu A, Mercan N, Turkoglu H, Gulcan S (2006) Free-radical scavenging capacity and antimicrobial activity of wild edible mushroom from Turkey. Afr J Biotechnol 5:1924–1928

Giler MRS, Roberty MMR, Montes YMG, Cool MEI, Vera MRL (2017) Evaluación del crecimiento micelial y productivo de *Pleurotus sapidus* a nivel in vitro y sobre residuo de maíz (Zea mays). La Técnica: Revista de las Agrociencias:16–24. ISSN 2477-8982

Gómez-Flores LDJ, Martínez-Ruiz NDR, Enríquez-Anchondo ID, Garza-Ocañas F, Nájera-Medellín JA, Quiñónez-Martínez M (2019) Análisis proximal y de composición mineral de cuatro especies de hongos ectomicorrícicos silvestres de la Sierra Tarahumara de Chihuahua. TIP. Revista especializada en ciencias químico-biológicas, vol. 22

Guzmán G (2008) Análisis de los estudios sobre los Macromycetes de México. Rev Mex Micol 28: 17–24

Gysling-Caselli J, Aguirre-Alvarado JJ, Casanova del Río K, Chung-Guin-Po P (2005) Estudio de mercado: Hongos silvestres comestibles. Instituto Forestal, Concepción

Herrera AE (2022) Situación actual de los bosques de Chihuahua. Madera Bosques 8(1):3–18

Instituto Nacional de Estadística y Geografía (INEGI) (En línea) (2013) Dirección de internet: http://www.inegi.org.mx

Kaewgrajang T, Sangwanit U, Iwase K, Kodama M, Yamato M (2013) Effects of ectomycorrhizal fungus *Astraeus odoratus* on *Dipterocarpus alatus* seedlings. J Trop For Sci:200–205

León-Guzmán MF, Silva I, López MG (1997) Proximate chemical composition, free amino acid contents, and free fatty acids contents of some wild edible mushrooms from Queretaro, México. J Agric Food Chem 45:4329–4332

López JMS, Saldaña MCM, Camacho RR, Morales RMC, Cardoso MLS, Lara FG (2017) Capacidad antioxidante y caracterización fitoquímica de extractos etanólicos de huitlacoche (*Ustilago maydis-Zea mays*) crudo y cocido. Rev Mex Cienc Farm 48(3):37–47

Martínez-Carrera D, Morales P, Sobal M, Bonilla M, Martínez W, Mayett Y (2012) Los hongos comestibles, funcionales y medicinales: su contribución al desarrollo de las cadenas agroalimentarias y la seguridad alimentaria en México. In: A. Menchaca-Rocha (Presidencia). Memorias Reunión General de la Academia Mexicana de Ciencias: Ciencia y Humanismo (Agrociencias). Academia Mexicana de Ciencias, Mexico, DF

Martínez-Carrera D, Sobal M, Morales P, Martínez W, Martínez M, Mayett Y (2004) Los Hongos Comestibles: Propiedades Nutricionales, Medicinales, y su Contribución a la Alimentación Mexicana. El Shiitake. COLPOS-BUAPUPAEP-IMINAP, Puebla

Martínez-Escobedo NA, Vázquez-González FJ, Valero-Galván J, Álvarez-Parrilla E, Garza-Ocañas F, Najera-Medellin JA, Quiñónez-Martínez M (2021) Antimicrobial activity, phenolic compounds content, and antioxidant capacity of four edible macromycete fungi from Chihuahua, Mexico. TIP. Rev Espec Cienc Quím -Biol 24

Martín-Pinto P, Pajares J, Díez J (2006) In vitro effects of four ectomycorrhizal fungi, *Boletus edulis, Rhizopogon roseolus, Laccaria laccata* and *Lactarius deliciosus* on *Fusarium* damping off in *Pinus nigra* seedlings. New For 32(3):323–334

Naranjo N, Andrade S, Herrera J, Ávila J, Almaraz N Gurrola N (2013) Análisis Proximal de seis especies de hongos silvestres comestibles en la región de El Salto, Pueblo Nuevo, Durango. (En línea). Dirección de interner: https://smbb.mx/congresos%20smbb/veracruz01/TRABAJOS/AREA_XIII/CXIII-60.pdf. Fecha de acceso: 26/04/2022

Pérez-Moreno J, Lorenzana-Fernández A, Carrasco-Hernández V, Yescas-Pérez A (2010) Los Hongos comestibles silvestres. Colegio de Postgraduados, campus Montecillo, Texcoco

Pompa-González, A., Aguirre-Acosta, E, Encalada-Olivas, AV, De Anda-Jauregui, A, Cifuentes-Blanco, J, Valenzuela-Garza R (2011) Los macromicetos del jardín botánico de ECOSUR "Dr. Alfredo Barrera Marin" Puerto Morelos, Quintana Roo. CONABIO, No. 6

Pontón J (2008) La pared celular de los hongos y el mecanismo de acción de la anidulafungina. Rev Iberoam Micol 25(2):78–82

Quiñonez M, Garza F, Sosa M, Lebgue T, Lavin P, Bernal S (2008) Indices de diversidad y similitud de hongos ectomicorrizogenos en bosques de Bocoyna, vol 33. Rev Ciencia Forestal en México, Chihuahua

Quiñonez-Martínez M, Garza-Ocañas F, Anguiano-Filio S, Chacón-Ramos V, Bernal-Carrillo S (2010) Diversidad de hongos comestibles en los bosques de Bocoyna y Urique del estado de Chihuahua". Ciencia en la frontera: revista de ciencia y tecnología de la UACJ. Vol. VII, pp 29–34

Quiñónez-Martínez M, Lebgue Keleng T, Covarrubias A, Cerecedo MS (2008) Diversidad de la vegetación en cuatro comunidades forestales con grado de disturbio en el municipio de Bocoyna, Chihuahua. Ciencia en la frontera 141

Quiñónez-Martínez M, Ruan-Soto F, Aguilar-Moreno IE, Garza-Ocañas F, Lebgue-Keleng T, Lavín-Murcio PA, Enríquez-Anchondo ID (2014) Knowledge and use of edible mushrooms in two municipalities of the Sierra Tarahumara, Chihuahua, Mexico. J Ethnobiol Ethnomed 10(1):1–13

Rentería-Chávez MC, Pérez-Moreno J, Cetina-Alcalá VM, Ferrera-Cerrato R, Xoconostle-Cázares B (2017) Transferencia de nutrientes y crecimiento de *Pinus greggii* Engelm. inoculado con hongos comestibles ectomicorrícicos en dos sustratos. Rev Arg Microbiol 49(1):93–104

Ríos-Hurtado A, Musquera LH, Torres MG, Hinestroza LI (2005) Caracterización bromatológica y compuestos bioactivos de la seta *Auricularia auricula.* Revista Institucional. Universidad Tecnológica del Choco D.L.C. 22:45–48

Ruan-Soto F, Mariaca R (2012) El mundo de los hongos silvestres comestibles. Ecofronteras:8–11. Available at: https://ecosur.repositorioinstitucional.mx/jspui/bitstream/1017/2351/1/51049_Documento.pdf

Sánchez JE, Mata G (2012) Hongos comestibles y medicinales en Iberoamérica. Limusa, México, pp 145–154

Seidl-Seiboth V, Ihrmark K, Druzhinina I, Karlsson M (2014) Molecular evolution of trichoderma chitinases. Biotechnol Biol Trichoderma:67–78. https://doi.org/10.1016/b978-0-444-59576-8.00005-9
Walker GM, White NA (2017) Introduction to fungal physiology. In: Fungi: biology and applications, pp 1–35
Wang P, Zhang Y, Mi F, Tang X, He X, Cao Y, y J. Xu. (2015) Recent advances in population genetics of ectomycorrhizal mushrooms *Russula* spp. Mycology 6(2):110–120
Wasser SP (2002) Medicinal mushrooms as a source of antitumor and immunomodulating polysaccharides. Appl Microbiol Biotechnol 60(3):258–274

Origin and Cultural Impact of Wild Chilli Pepper (*Capsicum annuum L. var. glabriusculum*) in Northeastern Mexico

Horacio Villalón-Mendoza, Nelson Manzanares-Miranda, Moisés Ramìrez-Meráz, Yesenia Marisol Mejorado-Martínez, and Juan Manuel Soto-Ramos

Abstract It is well known the important role that Mesoamerican cultures had in the domestication of wild chilli pepper *Capsicum annuum* in Mexico. The wild chili or "wild chili" (*C. annuum* L. var. *glabriusculum*), due to their role in the historical process of regional development, represents a natural and cultural resource, typical of northeastern Mexico. The objective of this research was to understand and analyze the impact of the historical process in the preference and cultural roots of the "wild chili" in the population of the Northeast of Mexico, to understand the strengthening of the northeast Mexico lifestyle. It used surveys and results showed that they were statistically significant ($p < 0.05$). The results showed that 74% of the population in the study area consumes almost all year-round fruits of wild chili, using techniques of production and preservation of fruit for it. However, there is a need for more protection of the natural resources of the northeastern Mexico, looking for strategies that generate ecological consciousness in the collectors for the management and sustainability of the species. The price of a kilogram of wild chili, is related to the number of fruits collected per day by the people, which in turn depends on the climatic conditions of the season of the year concerned, which can reach up to 85 US dollars per kg of fresh fruit in the month of April. The participation of women in productive processes or value chain of wild chili in the rural area is difficult due to cultural reasons prevailing in northeastern of Mexico. This variety of wild chili has a great touristic and gastronomic potential in the Northeast Mexico.

Keywords Wild chili · Gastronomic culture · Uses and customs · Northeastern Mexico

---

H. Villalón-Mendoza (✉) · M. Ramìrez-Meráz · Y. M. Mejorado-Martínez · J. M. Soto-Ramos
Facultad de Ciencias Forestales, Universidad Autónoma de Nuevo León, Linares, Nuevo León, Mexico

N. Manzanares-Miranda
Centro de Investigación en Producción Agropecuaria, Universidad Autónoma de Nuevo León, Linares, Nuevo León, Mexico

F. Garza-Ocañas (ed.), *Sustainable Management of Natural Resources*, Earth and Environmental Sciences Library, https://doi.org/10.1007/978-3-031-33394-1_10

1 Introduction

Currently the economic problems worldwide have grown considerably in recent years. Seriously harming all rural areas of the world and the country. Mexico is one of the three countries, along with Guatemala and Venezuela, where poverty has grown the most in the world. Currently there are several alternatives with which the most vulnerable population, rural areas, can increase their economic level, taking sustainable advantage of the resources that surround them and implementing small local businesses. Being the chili (*Capsicum* spp.) one of the most important crops in Mexico and in the world; the economically affected population can resort to this resource as a source of income for their family. The collection and extraction of fruits from wild relatives of this genus has been a way of taking advantage of natural resources and that to date subsists in many rural communities of the world. This traditional form of exploitation, in addition to representing additional income to the peasant family economy, is also the cultural expression of a long process of Human interaction with the environment, hence, in proposing alternatives for the use of natural resources, both ecological and economic aspects and socio-cultural aspects must be considered (Medina et al. 2003).

The genus *Capsicum* is native to the American continent, and chili peppers (wild and domesticated) were already used by various native peoples in pre-Columbian times. Chili peppers dispersed outside the continent after the arrival of Europeans and are now cultivated and consumed all over the world. The species that presents the greatest morphological diversity is *C. annuum*, particularly domesticated chilli peppers, with a wide spectrum of both shapes and colors of fruits, and in the level of pungency (Aguilar Meléndez 2019).

The wild chilli pepper is considered in rural areas of northeastern Mexico as a plant that generates a remuneration in critical times when the activities of the field are diminished, and the lack of employment becomes greater. Because the fruits, in addition, are products that are consumed fresh, dry, processed in canned with vinegar and in very different sauces, all of them with a very promising potential market.

In addition to economic retribution, the consumption of hot red chili peppers is associated with a 13% reduction in the immediate risk of death. Therefore, adding wild pepper in the daily diet of the population would reduce the absolute risk of mortality by 12%.

1.1 Origin

The genus *Capsicum*, in which chili is classified, belongs to a taxonomic subgroup of the Solanaceae family. This genus includes at least 35 species native to tropical America. For several years there has been much interest in studying the genetic diversity of the chili, mainly among the different fruits selected locally and that are usually designated as varieties, some domesticated native varieties and their wild relatives have been explored using various methodologies. Studies showed that genetic variability between Varieties of *Capsicum annuum* with sweet and large fruits is very restricted and it has been suggested that the loss of diversity during the transition from wild to cultivated populations occurred even in cohabitation areas. The low levels of diversity in the main gene reserve of chili is one of the reasons that have limited genetic improvement, another has been the lack of a reference genome (Aguilar Meléndez 2019). All species of the genus *Capsicum* are native to America. The pre-Columbian distribution of this genus probably extended from the southernmost edge of the United States to the warm temperate zone of southern South America (Montes Hernández 2019). Regarding the origin of this genus, one of the most accepted hypotheses about the place and mode of evolution of *Capsicum* species suggest that an important portion of the genus *Capsicum* originated in a "core area" in south-central Bolivia, with subsequent migration to the Andes and lowlands of the Amazon accompanied by radiation and speciation (McLeod et al. 1982). On the other hand, García (2018), mentions in relation to the genome of the chili (genus Capsicum), that the evolutionary history of species, which reflects their kinship relationships, is usually represented in the form of a branched scheme called a cladogram, familiarly referred to as a phylogenetic tree.

The 35 species recognized in the genus were segregated according to their phylogenetic affinities in 11 clades. In general, there is congruence between the clades and the geographical distribution of the species that make them up, that is, it is possible to observe groups of species more related to each other that are found in nearby and /or overlapping areas (Aguilar Meléndez 2019). The group of purple flowers (*C. eximium*) would have migrated to the highlands of the Andes, with the consequent directional selection that would have given rise to *C. cardenasii* by founding effect and *C. pubescens* as a domesticated species. The group of white flowers would have migrated out of the nuclear area through the Mizque River, which pours into tributaries of the Amazon. The flow of these waters is through the lowlands of tropical Bolivia and the Amazon basin. The ancestor that originated the group of white flowers gave rise to *C. baccatum* in the relatively dry area of southern Bolivia; the wild form would continue to migrate through the river system and in the humid Amazon basin would give rise to the wild progenitor of the *C. complex annuum*.

The subsequent domestication process seems plausible to have occurred independently in several areas, employing different wild species. It is also possible that, after the initial domestication of one species, the stimulus diffused by trying to grow other wild species in different areas. It is currently believed that *C. pubescens* and *C. baccatum* were domesticated in Bolivia in adjacent areas. The *C. annuum* complex was domesticated at least twice, a C. *annuum* type in Mexico and a *C. chinense* type in the Amazon (Pickersgill 1969). The findings of archaeological remains of these species complement the previous proposal. Although it has been difficult to determine exactly which species the various reminiscences found refer to.

1.2 Genus Capsicum

The genus *Capsicum* belongs to the Solanaceae family, consisting of 31 species (Pickersgill 1984). This genus is naturally distributed throughout much of the American continent and hence emerged the cultivated forms that are now present in most of the world (Table 1).

Capsicum is home to 5 domesticated species: *C. annuum* L., *C. frutescens* L., *C. chinense* Jacq., *C. pubescens* R. & P. (Campos 2018). In Mexico there are four wild species: *C. annuum* var. *glabriusculum* (Dunal) Heiser & Pickersgill, *C. frutescens* L., *C. ciliatum* L. and *C. lanceolatum* (Greenm.) Morton et Standl. The first two species are extensively used by the inhabitants of the New World (Pickersgill 1984).

The species of *C. annuum* L. is the most important today as it includes many economically important varieties grown around the world. This species includes "wild" chili peppers (*C. annuum* L. var. *glabriusculum* (Dunal) Heiser & Pickersgill) and domesticated chili peppers (*C. annuum* var. *annuum* L.). *Capsicum annuum* L. var. *glabriusculum* (Dunal) Hein ser & Pickersgill, is the variety managed for several centuries in Mexico and that gave rise to most of the commercial varieties that are grown and used around the world (Aguilar-Meléndez et al. 2009), that is, it is the wild ancestor of all cultivated varieties of *C. annuum* L. var. *annuum* L., which includes among others the Chilli peppers: Poblanos, Serranos, Puya, Guajillo, Ancho and Morrones.

Table 1 Morphological characteristics distinguishing domestic *Capsicum* species and their nearest wild progenitors (Modified by Eshbaugh and summarized by Hernández Verdugo 1999)

Species	Chalice toothed	Construction	Corola: color (size and shape)	No. of flowers	Seed color and shape
C. pubescens	Yes	No	Purple	1 (2–3)	rough black
C. cardenasii	Yes	No	Purple	2–3	yellow smooth
C. eximium	Yes	No	Magenta to white, stains of yellow -greensish	2–3	yellow smooth
C. baccatum var. *pendulum*	Yes	No	White, stains of yellow -greensish	1	yellow smooth
C. baccatum var. *baccatum*	Yes	No	White, with spots of yellow -greensish	2–3	yellow smooth
C. annum var. *annum*	Yes	No	White, large	1	yellow smooth
C. annum var. *glabrisculum*	Yes	No	White, small	1 (2–3)	yellow smooth

1.3 *The Genus* Capsicum *in Mexico*

The genus *Capsicum* is characterized by having accumulation of capsaicin, a component responsible for the characteristic spicy flavor of some chili peppers. This component is exclusive to the genus *Capsicum*, which makes it very interesting for Mexican culture (Aguilar Meléndez 2019).

In terms of market, in this genus several terms are used, relative to the degree of maturity, which would be the green and dry types; on the other hand, due to their content of spiciness, they are considered, spicy and sweet or peppers. The denomination of green chili is for natural or processed fruits, most of these are used in that state and when dehydrated they pass to the group of dry, such is the case of ancho, mulatto, guajillo chilli pepper, among many others. The green color of the fruits is due to the high amounts of accumulated chlorophyll, the ripe fruits become red or yellow due to pigments (lycopercysin, xanthophyll and carotene). For its part, the generic term "secos = dried", is used for many chilli pepper that are left to ripen and dry or dehydrate, the already dried limes are widely used in Mexican cuisine, mainly intended for the artisanal industry of mole; some other chili peppers besides dried are smoked such as chipotle and morita chili.

In Mexico, chilli peppers are among the most important vegetables, with a sown area that fluctuates between 150 and 170 thousand hectares and a production value that exceeds 8 billion pesos. Worldwide, it has become the main spice with a production volume of more than 22 million tons.

In the last decade, the production of chili presents an accelerated growth, with an increase of more than 10% per year, due to the increase in its use in the food, medicine, cosmetology, repellents and dyes industries, among many others (Ramírez Meraz 2018).

On the other hand, the labor required in its production was estimated at an average of 120 to 150 daily wages per hectare (Laborde-Cansino and Pozo-Campodónico 1982), which is very likely to have increased due to the diversification of production systems, such as those of protected agriculture (greenhouses and shade house).

Mexico ranks second in the world production of green chilli pepper, representing 8%, out of a total of 24,822,167 metric tons (MT), of which China is recorded as the first producer with 57% of the total. On the other hand, in the dry chilli pepper Mexico is in tenth place with 2% of the total produced in the world, which is 2,613,124 MT, with India being the main producer with 46% of the total production (Faostat 2009).

In Mexico there are more than 40 varieties of chili peppers, perhaps there is no other place in the world where the interest in *Capsicum* is stronger than in Mexico, a country that has always played an essential role in the cultural dissemination of chili peppers, their genetic resources, their culinary preparations, and their associated traditional knowledge. These flavors continue to be part of the identity of the diverse cultures in Mexico and are very important to the biocultural heritage of the entire nation.

Chili is a multifaceted resource, it is used as a condiment, vegetable, and dye in the kitchens of the world, and also has different industrial uses that generate strong foreign exchange. And although it is cultivated in different parts of the planet, the American continent has its center of origin and is where its greatest morphological diversity is found (Campos 2018).

1.4 History

1.4.1 Potential

Obtaining *Capsicum annuum* var. *glabriusculum* through agroforestry systems is an excellent alternative for better production in northeastern Mexico. The interest in this type of system is due to the need to find better options for the problems of low production and land degradation in the tropics.

A society cannot advance by developing only the socio-economic and technological environment and forget about its natural resources, since the development of the human being depends on its environmental environment.

It is known the important role that Mesoamerican cultures played in the domestication of the *Capsicum annuum* chilli pepper in Mexico. Evidence of this is the great variability of cultivated forms that are used in the country and that thanks to the diversity of agroecological environments and pre-Columbian cultures, offer us a wide range of shapes, colors, aromas, flavors, and sizes that constitute a valuable contribution of Mexico to world gastronomy (Pozo and Ramírez 2003).

Mexico is known worldwide for corn, beans and especially for its condiments where chili and its different varieties stand out, marking a per capita consumption of approximately 50 g per day of these three products. Chili is important in Mexican culture its consumption dates to pre-Hispanic times and is currently consumed in all strata of Mexican society.

Chilli pepper has a long cultural tradition in Mexico. Archaeological remains of this crop have been found in the Tehuacán Valley, Pue., dated between 7000 and 5000 BC. Although it is estimated that it may have been the first domesticated crop in Mesoamerica; It can be said that it has been an indispensable ingredient in the food of Mexicans for thousands of years.

The diversity of flavors and degrees of spiciness that different types of chilies can provide are evident when Bernardino de Sahagún describes in detail the great variety of chilli pepper that were offered in the markets, list of stews based on chili and sauces prepared for sale.

Its use dates to pre-Columbian times, where its primary use was as a condiment, but also the different types of chili peppers played an important role as a source of vitamin C in different American cultures (Eshbaugh 1970). In addition, of several uses that our ancestors gave it as medicine, punishment, currency, tribute material, etc. (Long-Solís 1986).

In addition to the tributary aspect there is much other evidence which by the nature of this writing are not presented, but which clearly indicate that chili is and will be a cultural constant through the different eras of the history of Mexico, covering all social strata.

The species *Capsicum annuum* L. is one of the most diverse in terms of shapes and sizes, according to Sahugún (1961) in his Florentine Code left testimony of the things of New Spain among them regarding the chili which had several denominations according to the great variety of chilli pepper cultivated. This has given rise to the various dishes that can be found in the different regions of the country. The world consumption of chilli pepper stands out for the constant annual increase. It is estimated that 25% of the world's population consumes some type of chili daily.

The wild chilli pepper (wild), represents a cultural value in the country, and very well recognized in the northeast of Mexico, so the planting of backyard and the collections in the field of this product, are a traditional practice, which greatly support the family economy, also promoting the physical and mental health of those who make it Villalón et al. 2010).

The chili (*Capsicum annuum* L.) represents an important chapter in the history and culture of Mexico. Its consumption in its various variant's dates from pre-Hispanic times and is currently rooted in all socioeconomic strata of the country.

The state of Nuevo León is located at the Northeast of Mexico, where the wild chilli pepper (wild) represents a genetic resource of great ecological importance and socioeconomic value. Interest has been aroused in carrying out research activities that will lead to improving the situation of this resource for this region of Mexico and that reduces the risk of losing the plant genetic resources that, although depleted, are still available.

This wild chilli pepper is collected by hand by the producers of the field, but they need knowledge about the conservation of their product, to have this product in different seasons of the year to obtain better prices (Medina et al. 2003).

One of the activities of great importance to be carried out in scientific research is the conservation of the products of the wild chilli pepper, not only for the conservation itself, but also for the added value that can be given to this resource and its potential socioeconomic impact that it represents for the rural communities of Northeast Mexico.

Chili has been part of Mexican culture since pre-Columbian times. The wild chili (*Capsicum annuum* L. var. *glabriusculum*) is small and widely distributed in northeastern Mexico, where it plays an important role in the cuisine and economy of rural areas (Villalón et al. 2010).

The northeast of Mexico is framed in the context of arid and semi-arid areas. In this region predominates the vegetation typified as thorny scrub Tamaulipeco, which, has meaning an important source of resources for the development of rural communities that depend directly on obtaining fodder to feed the cattle herd; alternative source of energy using firewood and charcoal; wood for rural constructions; shelves for fences; and medicinal herbs.

Historically, the forms of extraction of resources from their natural habitat, in addition to representing additional income to the peasant family economy, is also the cultural expression of a long process of interaction of man with the environment.

The wild chilli pepper represents a natural and cultural resource typical of northeastern Mexico. This plant as a component of the submontane scrub of the region, is perennial in tropical and subtropical areas. The wild chilli pepper represents for the inhabitants of the region a food, medicinal and income-generating source (Medina Martìnez, 2007).

As for the activity of harvesting wild chilli pepper, it involves social groups with a high degree of marginalization and little organizational integration for production (Medina et al. 2003).

The wild chili (wild or from the countryside) of northeastern Mexico during the time of greatest supply comes to displace other types of chilli pepper for its pleasant flavor and degree of pungency, in addition, it does not irritate the digestive system.

The wild chilli pepper in northeastern Mexico receives many local or regional denominations is popularly known as "chilli pepper wild", "chilli pepper del campo", chilli pepper del monte", among others, which represents an important chapter in the history, culture, and tradition of Mexico, since its consumption dates to pre-Hispanic times and is considered the closest ancestor to *the Capsicum annuum chilli pepper* that are grown around the world.

They are small oval fruits of 5–7 mm in diameter that grow with solitary peduncles, creamy white flowers, their fruits are from green changing to orange and bright red with the ripening of this, they contain an average of 18 seeds per fruit.

The wild chilli pepper of northeastern Mexico is a species with economic potential for the rural environment and is important for the tradition and culinary culture of Northeast Mexico, since this wild species is strongly rooted in its typical gastronomy, since it intervenes in the diet of the Norestenses in various ways; either by the consumption of its fresh or dried fruit, in salads, sauces, pickles, moles, mixtures to marinate meats, etc. Chili becomes finished products for sale and has economic potential in the national and international market (Almanza 1993).

From the ecological point of view, the current management of biotic resources in northeastern Mexico could affect the future permanence of wild chilli pepper and reduce its genetic potential (Villalón et al. 2009). This species in natural areas depends entirely on ecological conditions, such as soil, altitude, rainfall, potential evaporation, and temperatures.

From the social point of view, it is estimated that 15% of the rural population of northeastern Mexico is dedicated to collecting wild chilli pepper in the months of September to December, during this period 60% of the income obtained by these communities is through the collection and sale of the product when there is a surplus. The remaining 40% is made up of other agricultural activities. This activity is carried out mostly by women and children, occasionally the producer enters the process.

From the economic point of view, its commercialization comes mainly from collections of wild populations, which is mostly harvested without a sustainable management plan, thus generating a deterioration of the ecosystem and natural germplasm banks, coupled with the difficulty presented by the germination of the seed. That way the prices of wild chilli pepper in the market depend mainly on the prevailing weather conditions in each season.

This wild species from northeastern Mexico is a very spicy chili, but its characteristic flavor is highly appreciated, so much so that it pays from 30 to more than 340 times the value of cultivated commercial chilli pepper, such as serrano or jalapeño.

The demand for this species and its socioeconomic importance as a natural resource, mainly in rural areas, make it a species with potential for its production, becoming an alternative for in situ conservation and productive ex situ in a sustainable way. The chili intervenes in the daily diet of Mexicans in various presentations, whether in green, dry, powder, pickle, sauces, salads, moles, stuffed chilli pepper, sweets, and others (Rodríguez del Bosque et al. 2003).

The chili "wild" or "del monte" (*C. annuum*, L. var. *glabriusculum* Dierb.), considered as the ancestor of all forms of chili peppers currently known within this species (jalapeño, serrano, ancho, pasilla, guajillo, de árbol, etc.), is widely distributed in the wild in Mexico, mainly in the lowlands. The fruit of this chili is appreciated and very well valued, that even during the time of greatest production in the natural vegetation, its price exceeds those of other types of chilies, due to its pleasant flavor and degree of pungency; In addition, it does not irritate the digestive system due to the more alkaline pH of its capsaicin, than most chili peppers.

The acceptance of chili peppers in the international context is not only limited to the organoleptic aspect of flavorings as a special, but also for its wide attributes that are used in cosmetology, pharmacology, traditional and modern medicine, pigments for food and industrial, refreshing drinks, in the elaboration of aerosols for self-defense and others. This has given rise to the formation of a thriving national and international industry that ranges from the production of homemade products to large transnational consortia (Pozo and Ramírez 2003).

The Northeast of Mexico is framed in the context of arid and semi-arid areas. In this region predominates the vegetation typified as thorny scrub Tamaulipeco, which has meaning an important source of resources for the development of rural communities that depend directly on obtaining fodder to feed the cattle herd; alternative source of energy using firewood and charcoal; wood for rural constructions; shelves for fences; and medicinal herbs.

As for the activity of harvesting wild chili, it involves social groups with a high degree of marginalization and little organizational integration for production (Medina et al. 2003). The demand for wild pepper, in addition to its socioeconomic importance as a natural resource in rural areas, is its potential as a new productive option in northeastern Mexico (Rodríguez del Bosque et al. 2003).

Cantú and Villalón (2013), mention as a potential for the sustainable use of this species typical of northeastern Mexico that, it is possible to innovate the sustainable management of wild chilli pepper through rural tourism, since it represents one of the natural tourist resources of the rural area of northeastern Mexico, turning out to be a productive alternative for sustainable development, which to date has not been properly exploited.

In this regard, Cantú and Villalón (2013) mention that, according to the OAS, studies carried out to know the impact of tourism on local communities have found that a tourism program based on adequate studies, well planned and managed, which also considers the local, natural, and cultural environment, has good possibilities of improving the local economy and the quality of life of local residents. Some of the economic benefits include increased income, the development of products and services, and improved infrastructure, as well as opportunities to link with other services and products from other sectors of the economy. If the approach is right, it can also promote community pride and result in a better quality of life on site (OAS 1997).

Likewise, Boullón (2005) divides tourism and recreation activities into five categories:

1. Recreation
2. Cultural visits
3. Visits to natural sites
4. Sports
5. Attendance at scheduled events

Therefore, cultural visits, which at traditional convenience ensues; they refer to activities based in museums, artistic monuments, churches, archaeological ruins, historical sites, civil works of the past, traditional fairs and markets, handicrafts, religious manifestations and popular beliefs, visits to ethnic groups or expressions and popular architecture and attendance at scientific and technical centers and mining, agricultural or industrial exploitations (Boullón 2005).

About the tourist potential of Chilli pepper Wild

Analyzing all the above, it is pertinent to design the structure of a program of recreation activities where the protagonist is Chilli pepper Wild, as a tourist resource.

The recreation product: Refers to the activities to be developed in the producing region of Chilli pepper Wild:

1. Sowing the seed
2. Crop care
3. Harvest or collection of the fruit
4. Product development

Food and transport will be solved by the equipment and common services of the rural area.

The physical environment. Based on the fact that the rural area is where the producing regions of Chilli pepper Wild are located, in which recreation activities will have to be developed, it is determined that the area corresponds to proximity tourism (Ávila Aldapa 2007), that is, to short trips that can prolong the stay of 1 to 2 nights in the nearest con urban area, since the route to the rural area is integrated with the route of the municipality.

Structure of the program. The tourist in the producing region will be the protagonist during the tourist recreation tour in the rural area, mainly between the months of August-December, which is the main harvest time. The rest of the year recreation activities will focus only on the elaboration of products, emphasizing the months of March to June during planting and early production.

Composition of the program. The demographic level in both areas (rural and con urban) is not dense, for example, in Nuevo León 5% of the population is rural, in Linares the population oscillates in 70,000 inhabitants, therefore, approximately 3500 people live in its 76 ejidos. This is very interesting, since for the inhabitants of the cities it can be very interesting to practice rural tourism.

Regarding the receptive tourism segment, a study carried out by SECTUR in 2002, on the behavior of 1500 tourists in Mexico, Guadalajara, and Monterrey, in proximity tourism (Ávila Aldapa 2007). People from northeastern México decide days in advance this type of trips to rest and have fun; spending between 4001 and 8000 pesos, highlighting the fact that the experiences of friends, family or co-workers influence and motivate to choose the weekend destination or round trip. In this way, a promising panorama is presented in terms of the demand that the Northeast of Mexico in high season and weekends can present. Turning very much into account, that the quality of care in the service will be the best recommendation. According to a statistic published in January 2011, in the newspaper El Norte, the citizens of Nuevo León will choose 47% to vacation in the same State.

Strategy. The producing regions can be organized to work as a team with the municipal tourist offices or with a receptive travel agency in the city, this to carry out quality organized tours.

Result of the program. The benefits that the administration and execution of the program for wild chilli pepper producers will bring are:

1. Economic spillover
2. Regional tourism projection
3. Recognition in the agricultural and tourism industry
4. Increase knowledge about the potential of natural resources and their cultural relationship

While for tourists the benefits are:

1. Cultural
2. Training
3. Enjoy nature
4. Break with routine
5. Living longer with your family
6. Know other regions of the entity
7. Educational knowledge

In the diversity is the success of the competitiveness of a tourist destination, therefore, considering that the local and rural destinations of the northeast of Mexico receive in a greater percentage the receptive tourism, diversifying the offer and promoting it, will attract their interest and awaken their need to go to that tourist destination.

Therefore, as Molina (2009) points out, the achievement of higher levels of competitiveness of the sector's offers, from emphasizing the uniqueness of the

Fig. 1 Mature (red) and inmature (green) wild chilli peppers

strategies and projects, and the increase in the effective impact of the benefits of tourism in the spaces of the resident communities where the activity is manifested, are fundamental concepts to promote tourism that affects local living conditions.

The ecological, socio cultural and socioeconomic potential of the wild chilli pepper makes it ideal for rural tourism activities, so they represent a sustainable and viable productive alternative for the inhabitants of the rural environment of northeastern Mexico.

In a study by Villalon-Mendoza et al. (2014) that aimed to know and analyze the impact and effect of the historical process on the preference and cultural roots of the wild chilli pepper "wild" (*Capsicum annuum* L. var. *glabriusculum*) in the population of northeastern Mexico, analyzing its role in strengthening the Northeastern lifestyle, found that the ways of managing the wild chilli pepper in northeastern Mexico range from backyard cultivation to its greenhouse cultivation, through the collection in the characteristic natural vegetation, predominating its natural production in the Tamaulipas scrub types (See in Fig. 1).

As can be seen in Fig. 2, 74% of the population of northeastern Mexico consumes wild chili fruits almost all year round, either because they grow it at home, store it dry, freeze it, or preserve it in sauces, brine, or pickle (Fig. 2).

In northeastern Mexico people collect the fruit in the "monte" regardless of the associated risks, and this due to the need for income for family subsistence, since it is known that the wild chilli pepper enters the food chain where there are also the typical birds consuming wild chili (*Mimus poliglottos* and *Toxostoma longirostre*) and rattlesnakes (*Crotalus atrox*), where the latter feeds on the first, and in turn the second of the fruits of the wild chili, which when going to harvest fruits i.e. pisca, can occur the encounter of the human being with the snakes, this is also mentioned in the research of Medina et al. (2000).

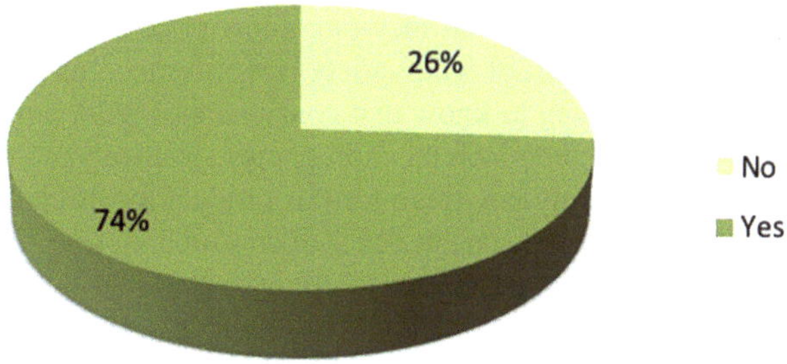

Fig. 2 Percentage of families that have wild chili and it is part of their typical diet all year round, in addition to considering it a delicacy to share with close and appreciated people

It was observed that 62% of the wild chilli pepper collectors harvest the fruit of the plant with everything and the cape. 38% of the collectors harvest the fruit by cutting bunches of branches and 34% sell their product, these activities that go against the sustainable management of the Natural populations as mentioned by Almanza (1993) may soon bring irreversible consequences to the loss of genetic potential of these populations.

Several activities were found as the most used to produce and maintain the wild chilli pepper throughout the year, it should be noted that many of the inhabitants do several of the actions, being the process of drying the fruits the most used with 24% of the population, drying it in the shade and a third of them drying it by hanging the branches with fruits down and on top of the kitchen stove where sometimes the smoke from the fire comes to cook. It was followed by the freezing of clean fruits, with 22% of people consuming and trying to have wild chili all year round. Next, the cultivation of plants in their gardens was located to harvest it directly from the plants and the conservation in vinegar (the pickle) and the preparation in sauces, with 18%, 17% and 17%, respectively. Finally, fresh consumption was found with only 2%.

A data of great importance obtained by Villalon-Mendoza et al. (2014) in their research was to know the causes of the behavior of the wild chilli pepper market, due to which, productive activities can be carried out aimed at satisfying the quality demands of this product and taking advantage of it in a sustainable way the value that this resource represents, which ranges from the economic to the sociocultural through the ecological and productive technician.

The observed results showed the willingness, of the inhabitants of northeastern Mexico, to pay a high price for the kilogram of wild chili, which predominates in quantities between \$3.60-\$7.10 US dollar with 35%, following with \$7.11-\$10.71 US dollar with 15%, being at the same time a not very high amount and with only 1% in the range of \$32.20-\$35.70 US dollar, the latter representing the highest price the population is willing to pay, even though it was mentioned by some respondents; know of people who have paid up to \$85 US dollar for each kg of fresh green wild chili and up to \$180 US dollar per kg of dried red fruit.

It was observed that the reasons why people are willing to pay a high economic price for the fruits of wild chili; resulting with greater predominance (38% of the people surveyed) the pleasant taste that the fruit of this product presents and with which they fully identify, followed by the low economy that people have with 14%

(referring to the fact that they pay a low price for lack of economic resources, but aware that the fruit of wild variety is worth much more to them).

It is interesting to know that consumers of wild variety pay more for wild chilli pepper than for commercial chilli pepper (jalapeño and serrano); 38% of consumers of the wild variety pay more for this product for its characteristic flavor and aroma, which make it irreplaceable (12% of consumers), pay it better for the work it costs to harvest it, including the risk of going to the field to harvest it, this coincides with what is mentioned by Villalón Mendoza et al. (2013), who claim that this species has very small fruits, so it is difficult to harvest and the amount of fruits harvested by hand per hour. With almost the same proportion (10% of consumers) they pay more for this product due to the quality of the product and the supply and demand. For its characteristic itching it was mentioned by 9% of the respondents and for considering it very healthy (more than the other commercial varieties) it was mentioned by 8%.

It is relevant to mention that 7% of the population of northeastern Mexico appreciates it and pays for being a regional product, because they also identify culturally with this product that has historically accompanied the first settlers of this important region of the country. Finally, other causes are; because they take advantage of the excellent market that this product presents and resell it, buying wholesale and selling retail (4%) and to give it added value (in the preparation of sauces, preserves, etc.) 2%. It was confirmed that the harvest is traditionally done by hand and the production can be sold for fresh consumption or carry out productive activities of transformation to give added value to the product, the latter activity being the one that brings better economic impacts to the countryside of northeastern Mexico. The problem encountered to carry out this operation to increase the added value, was that having to enter the woman of the family to the productive chain, this complicated things, since historically in the field it is governed by a family organization of patriarchy and the fear of the heads of family was that women will be empowered and thus lose their status in the family and in community, so they usually chose to make sales directly of the fresh product, where they (the male gender) have economic control of the productive system, regardless of whether to gain considerable economic potential (Fig. 3).

It was found that, in 2014, 74% of the population of northeastern Mexico consumes almost all year-round wild chilli pepper fruits, either because they grow it at home, store it dry, freeze it, or preserve it in sauces, brine or pickle. 78% of families that have wild chilli pepper at their homes, have it only in one type of presentation, while 14% had two types of presentation, with a lower percentage it was found that 5% with three types, 2% with five types, 1% with four. 24% of the inhabitants have wild chilli pepper in their home dried or dehydrated, while 22% keep it frozen, on the other hand, 17% keep it in processed products such as salsa and escabeche. It was found that people in northeastern Mexico offer more money for wild chilli pepper than for any other commercial chilli pepper, because they consider it to be a product of better flavor, quality, recognize the industriousness required for its production and consider it a regional resource with which they identify culturally. On average, in 2019, people in northeastern Mexico would pay about US$12.6 for a kilogram of wild chili, with a standard deviation of US$7.6, when serrano and

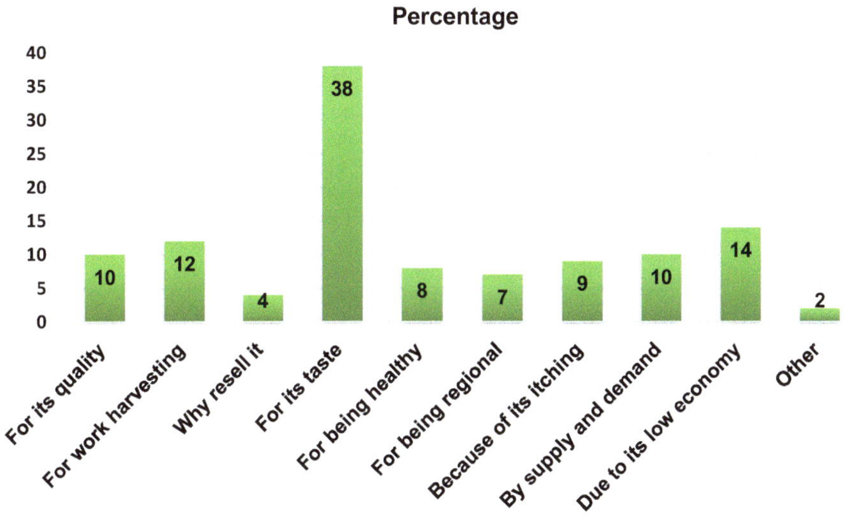

Fig. 3 People's reasons for preference of wild chili pepper

jalapeno pepper cost US$0.25. Greater protection of the natural resources of northeastern Mexico is required, seeking strategies that generate ecological awareness in the collectors and legislation for the sustainable management of the species.

References

Aguilar Meléndez A (2019) Los chiles silvestres que le dan sabor al mundo (Primera ed., págs. 26–40). Xalapa, Veracruz, México: Universidad Veracruzana

Aguilar-Meléndez A, Morrel PL, Roose ML, Seung-Chul K (2009) Genetic diversity and structure in semiwild and domesticated chilli pepper (*Capsicum annuum*, Solanaceae) from México. Am J Bot 96:1190–1202

Almanza FJG (1993) El chile silvestre (*Capsicum annuum* L. var. *aviculare* Dieb) Estudio etnobotánico, biología y productividad. Bachelor Thesis. Biology Faculty, Universidad Autónoma de Nuevo León, México. pp 32–72

Ávila Aldapa RM (2007) Turismo Cultural en México. Trillas, Mexico, D.F.

Boullón CR (2005) Las Actividades Turísticas y Recreacionales. Trillas, México, D.F.

Campos ER (2018). La diversidad genética de *Capsicum annuum* en México. En M. A. Araceli Aguilar Meléndez, Los chiles que le dan sabor al mundo (Primera ed., págs. 52–67). Xalapa, Veracruz, México: Universidad Veracruzana. Recuperado el abril de 2019

Cantú MAF, Villalón MH (2013) El valor cultural del chile piquín como recurso turístico para apoyar el desarrollo sustentable de Linares, Nuevo León, México. Capítulo 23 en Pérez-Mendoza, S., Bueno C., L.E. y Meza C., F. Gobiernos locales y desarrollo municipal. Análisis y propuestas para el desarrollo local en el marco de la globalización. Benemérita Universidad Autónoma de Puebla. Universidad de Costa Rica. Universidad Autónoma de Tamaulipas. Red de Estudios Municipales VIII Congreso Internacional de la Red-ε-Mun "Gobiernos Locales y Desarrollo Municipal", Tehuacán, Puebla, México, octubre 2013. Cuerpo Académico de

Economía Urbana y Regional, Facultad de Economía, BUAP. Edit. Impresos Proyección Grafica. Puebla, Puebla, México. 1ª. Ed. 23:433–441
Eshbaugh WH (1970) A biosystematic and evolutionary study of *Capsicum baccatum* (Solanaceae). Brittonia 22:31–43
FAOSTAT (2009) Production crops. Food and Agriculture Organization of the United Nations. http://faostat.fao.org/site/339/default.aspx. Accessed June 2009
García CC (2018) Breve historia evolutiva del género *Capsicum*. En M. A. Araceli Aguilar Meléndez, Los chiles que le dan sabor al mundo (Primera ed., págs. 26–40). Xalapa, Veracruz, México: Universidad Veracruzana. Recuperado el abril de 2019
Hernández Verdugo SPD (1999) Síntesis del conocimiento Taxonómico, origen y domesticación del género *Capsicum*. Boletín de la Sociedad Botánica de México:65–84
Laborde-Cansino JA, Pozo-Campodónico O (1982). Presente y pasado del chile en México. Publicación especial No. 85. lNIA-SARH, México
Long-Solís J (1986) *Capsicum* y cultura: la historia de chile. Fondo de Cultura Económica, México, p 203
McLeod, MJ, Guttman, SI, Eshbaugh WH, 1982. WH Econ Bot 36: 361. https://doi.org/10.1007/BF02862689
Medina MT, Villalón HM, Lara V, Gaona G, Trejo LH, Cardona AE (2000) El chile del Noreste de México. CONACYT, UAT y UANL. 37p
Medina MT Villalón HM, Rodríguez del Bosque LA, Pozo CO, Ramírez MM, Lara VM, Trejo HL, Cardona EA, Mora OA, Carreón A (2003) Estudio poblacional y manejo agroforestal de chile silvestre (*Capsicum annuum* L. var. *aviculare*) en el noreste de México. I.er Simposio Regional sobre chile silvestre: Avances de Investigación en Tecnología de Producción y Uso Racional del Recurso Silvestre. Río Bravo, Tam., México. pp 17–20
Medina Martínez TC (2007) Socioeconomía y ecología del chile en Tamaulipas, México. en Cuarta Convención Mundial del Chile. Querétaro, Qro. pp 276–279
Molina S (2009) Fundamentos del Nuevo Turismo. Trillas, México, D.F.
Montes Hernández, SPL (s.f.) Recopilación y análisis de la información existente de las especies del género *Capsicum* que crecen y se cultivan en México. INIFAP, Celaya. Recuperado el 2019, de. https://www.biodiversidad.gob.mx/genes/centrosOrigen/Capsicum/Proyecto/Proyecto%20Capsicum.pdf
OEA Sustentabilidad del turismo mediante la gestión de los recursos naturales y culturales, en XVII Congreso Interamericano del Turismo, <Doc.OEA/Ser.K 111.1>, abril de 1997
Pickersgill B (1969) The archeological record of chili peppers (*Capsicum* spp.) and the secuence of the domestication in Perú. Amer Antiq 34:54–61
Pickersgill B (1984) Migrations of chili peppers, Capsicum spp., in the Americas. pp 105–123 in D. Stone, (ed.). Pre-columbian plant migration. Papers of the Peabody Museum of Archaeology and Ethnology Vol. 76. Harvard University Press, Cambridge, Massachusetts, USA
Pozo CO, Ramírez MM (2003) Diversidad e importancia de los chiles silvestres. I.er Simposio Regional sobre Chile silvestre: A v a n c e s d e Investigación en Tecnología de Producción y Uso Racional del Recurso Silvestre. Río Bravo, Tam., México. pp 3–7
Ramírez Meraz M (2018) Mejoramiento genético de los chiles comerciales en México. En M. A. Araceli Aguilar Meléndez, Los chiles que le dan sabor al mundo (Primera ed., págs. 286–293). Xalapa, Veracruz, México: Universidad Veracruzana. Recuperado el abril de 2019
Rodríguez del Bosque LA, Ramírez MM, Pozo CO (2003) El cultivo del chile silvestre bajo diferentes sistemas de producción en el noreste de México. I.er Simposio Regional sobre Chile: Avances de Investigación en Tecnología de Producción y Uso Racional del Recurso Silvestre. Río Bravo, Tam., México. pp 8–14
Sahugún B (1961) Florentine Codex- The General History of The Tings of New Spain, traducido al inglés por AJO. Anderson y C Dibble, The School of American Research & The University of Utah, Santa Fe

Villalón M, Garza F, Soto R, Medina JM, Ramírez TM, López de LR, Carrillo PA (2009) Aprovechamiento y manejo del chile silvestre (*Capsicum annuum* L. var. *aviculare* Dierb.) en el Parque Cumbres de Monterrey. Memorias del VII Congreso Nacional Sobre Áreas Naturales Protegidas de México. San Luis Potosí. pp 76–81

Villalón MH, Méndez VE, Ramírez MM, Medina MTY, Garza OF (2010) Estatus del chile silvestre (*Capsicum annuum* L. var. *glabriusculum/aviculare*) en Nuevo León, México. Proceedings of Seventh World Pepper Convention. 13 al15 de junio. Aguascalientes, Ags., México

Villalón Mendoza H, Medina Martínez T Ramírez Meráz M (2013) Factores de calidad de la semilla de chile silvestre (*Capsicum annuum* L. var. *glabriusculum*). Quality factors of wild chilli pepper (*Capsicum annuum* L. var. *glabriusculum*) seeds. Revista Mexicana de Ciencias Forestales. Edit. INIFAP. Vol. 4. Núm 17 mayo-junio 2013. ISSN: 2007-1132. Pág. 182–187

Villalon-Mendoza H, Medina-Martínez T, Ramírez-Meráz M, Solís Urbina SE, Maiti R (2014). Factors influencing the price of chilli pepper (*Capsicum annuum* L. var. *glabriusculum*) of North-east Mexico. International Journal of Bio-resource and Stress Management. Edit. Puspa Publishing House. ISSN 0976-3988 OnLine 0976-4038 Volume & Issue: 5(1)

Diversity of Macrofungi in the Forest Ecosystems of the Cumbres National Park

Fortunato Garza-Ocañas, Miroslava Quiñónez Martínez, and Artemio Carrillo Parra

Abstract Preliminary progress on the identification of macromycete species associated with different vegetation types at the Cumbres National Park is presented. The results presented here indicate that so far it has been possible to identify 215 species out of more than 1300 species collected so far in the vegetation types of the park. The species are distributed as follows: The mixed oak-pine forests have 190 species followed by the pine forests with 145, the oak forests with 144, the coniferous forests with 101, the thorn scrubs with 93, the riparian forest with 50, the oak chaparral with 48, the grasslands with 16, the microphylic desert scrub with 14 and the rosetophile desert scrub with 9. Of the species studied 95 are mycorrhizal, 106 are saprotrophic and 14 are parasitic either to plants or fruits of another fungus. Many of the species grow in various vegetation types so in the overall count they are repeated. Distribution maps of the species were carried out and some are shown.

Keywords Mushrooms · Diversity · Cumbres National Park forest ecosystems

F. Garza-Ocañas (✉)
Laboratorio de Micología, Facultad de Ciencias Forestales Universidad Autónoma de Nuevo León, Linares, Nuevo León, Mexico

M. Q. Martínez
Instituto de Ciencias Biomédicas, Universidad Autónoma de Ciudad Juárez, Ciudad Juárez, Chihuahua, Mexico

A. C. Parra
Universidad Juárez del Estado de Durango, Instituto de Silvicultura e Industria de la Madera, Ciudad Universitaria, Durango, Durango, Mexico

F. Garza-Ocañas (ed.), *Sustainable Management of Natural Resources*, Earth and Environmental Sciences Library, https://doi.org/10.1007/978-3-031-33394-1_11

1 Introduction

The Cumbres de Monterrey Park is located at 8 municipalities that are: Allende, García, Montemorelos, Monterrey, Rayones, Santa Catarina, Santiago and San Pedro Garza García from the State of Nuevo Leon and has a total area of 177,395.98 hectares. This park is a protected natural area and provides important ecological and environmental services to the city of Monterrey. Visitors to the park produce a strong ecological pressure causing imbalance in forest ecosystems by fragmenting them, in addition to producing changes in land use, urban growth, clandestine logging and frequent forest fires among other factors cause a decrease in the diversity of organisms including fungi. Fungi are important for forests because of the multiple interactions and ecological relationships they have with the biotic component and the biotechnological and medicinal potential they have (Garza et al. 1985; Garza 1986; Guzmán 1998; García and Garza 2001; Marmolejo et al. 2013; Boa 2004). So far, the species previously studied by several authors have been reported in addition to those that are deposited in some of the diversity of macrofungal species associated with the vegetation types of the park.

2 Objective

To study the diversity and distribution of macromycetes species known and their distribution in the different types of vegetation of the PNCM.

3 Methodology

Collections of carpophores were made in the different types of vegetation of the PNCM for 30 years. Data were recorded such as location, date, altitude, vegetation, hosts and georeferenced at the collection points, as well as field photography of the species. With all the data obtained, a database of the species was elaborated in Excel. The location of the collection sites was carried out using functions of the ARCVIEW GIS 3.2 Geographic Information System. The database was stabilized and georeferenced and information from the National Forest Inventory 2015 and the topographic charts of the INEGI scale 1:50,000 and scale 1:250,000 were used. The location sites of the macromycetes were displayed on the types of vegetation of the State and on the topographic charts to prepare the distribution maps of these. Given the space constraints no maps are presented.

4 Results

In this study, 215 species of macro fungi of the more than 1200 present in the park in the vegetation types of the park are presented and the distribution of the species occurs as follows: The mixed oak-pine forests have 190 species followed by the pine forests with 145, those of oak with 144, those of conifers with 101, the Scrub "sensu lato" with 93, the riparian forest with 50, the chaparral of oaks with 48, the Grasslands with 16, the microphyllic desert scrub 14 and the rosetophyles desert scrub 9 (See Fig. 1 and Table 1). Many species grow in various types of vegetation by which they are repeated in the general count giving a cumulative sum of 810 (Figs. 2, 3, and 4).

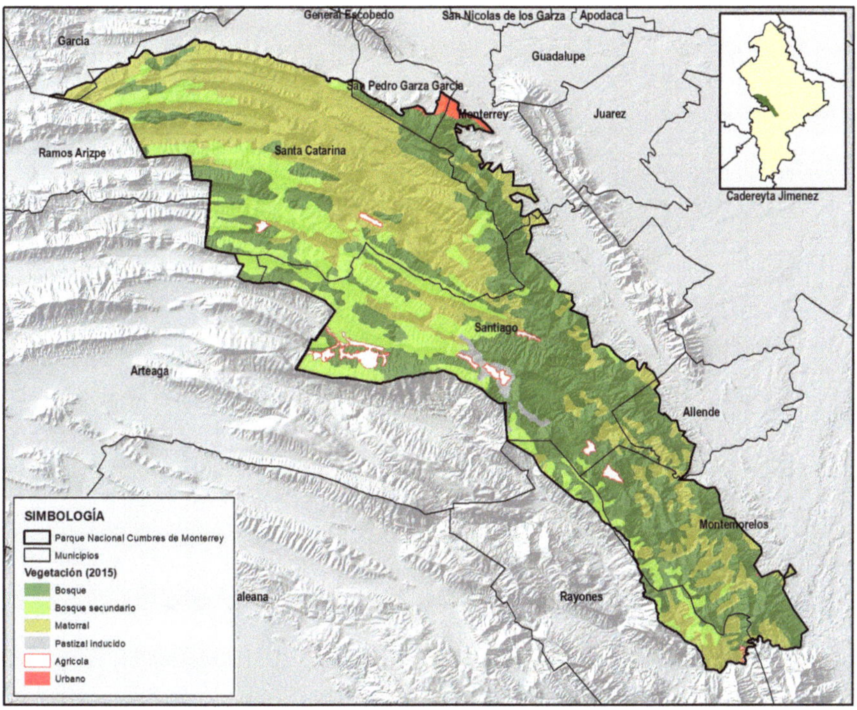

Fig. 1 Delimitation of Cumbres de Monterrey National Park and the main types of vegetation

Table 1 Species studied and their distribution by type of vegetation in the Cumbres de Monterrey National Park, Nuevo León

Species	Vegetation										
	1. BEP	*2. BP*	*3. BE*	*4. BC*	*5. M*	*6. R*	*7. ChE*	*8. P*	*9.MM*	*10.MR*	*Edibility*
Cordyceps militaris	*X*	*X*	*X*		*X*						*M*
Daldinia concentrica	*X*		*X*		*X*	*X*					
Xylaria polymorpha	*X*	*X*	*X*	*X*	*X*	*X*	*X*				
X. hypoxylon	*X*	*X*	*X*	*X*	*X*						
Leotia lubrica		*X*		*X*			*X*				
Chlorosplenium aeruginascens		*X*		*X*							
Helvella crispa	*X*	*X*		*X*			*X*				
Otidea onotica	*X*	*X*		*X*							
Aleuria aurantia	*X*	*X*	*X*	*X*	*X*	*X*					
Paxina acetabulum	*X*	*X*		*X*							
Humaria hemisphaerica	*X*	*X*	*X*	*X*							
Sarcoscypha coccinea	*X*	*X*	*X*			*X*					
Genea hispidula	*X*		*X*								
Tuber candidum	*X*		*X*		*X*						
Pachyploeus citrina	*X*		*X*		*X*						
P. virescens	*X*		*X*		*X*						
Elaphomyces muricatus	*X*		*X*		*X*						
Hypomyces lactifluorum	*X*	*X*	*X*	*X*							
Tremella mesenterica	*X*	*X*	*X*	*X*	*X*	*X*	*X*				
Phaeotrmella foliacea	*X*		*X*								
Tremella fuciformis	*X*		*X*		*X*		*X*				
Pseudohydnum gelatinosum	*X*	*X*	*X*		*X*						
Auricularia nigricans	*X*		*X*		*X*	*X*					
A. mesenterica	*X*		*X*		*X*	*X*					
Stereum complicatum	*X*		*X*		*X*	*X*					

S. ostrea	*X*	*X*	*X*		*X*	*X*					
Xylobolus subpileatus	*X*	*X*	*X*		*X*	*X*					
Thelephora caryophyllea	*X*		*X*			*X*	*X*				
Auriscalpium vulgare		*X*		*X*							
Sarcodon imbricatus	*X*		*X*								
H. repandum		*X*		*X*							*Y*
H. caeruleum	*X*	*X*	*X*								
Hericium erinaceum	*X*		*X*		*X*						*Y, M*
Daedalea quercina	*X*		*X*		*X*	*X*					
Trametes elegans	*X*		*X*		*X*	*X*					
Pycnoporus sanguineus	*X*		*X*		*X*	*X*					*M*
Trichaptum biforme	*X*	*X*		*X*							
Cantharellus cibarius	*X*	*X*	*X*	*X*	*X*	*X*					*Y*
Craterellus tubeiformis	*X*		*X*								*Y*
C. cinnabarinus	*X*		*X*								
C. Odoratus	*X*		*X*								*Y*
Craterellus cornucupioides	*X*		*X*								*Y*
Turbinellus floccosus	*X*	*X*	*X*								*Y*
Gomphus clavatus	*X*	*X*	*X*								
Schyzophyllum commune	*X*	*X*	*X*		*X*	*X*	*X*				
S. umbrinum	*X*		*X*		*X*	*X*					
Gloeostereum incarnatum	*X*	*X*	*X*		*X*		*X*				
Phlebia rufa	*X*		*X*		*X*	*X*					
Ganoderma applanatum	*X*		*X*			*X*					*M*
G. lucidum	*X*		*X*		*X*	*X*					*M*
G. lobatum	*X*		*X*		*X*						*M*
Humphreya coffeata					*X*						*M*

(continued)

Table 1 (continued)

Species	Vegetation										
	1. BEP	*2. BP*	*3. BE*	*4. BC*	*5. M*	*6. R*	*7. ChE*	*8. P*	*9.MM*	*10.MR*	*Edibility*
Albatrellus ellisii		*X*		*X*	*X*						
Coltricia focicola	*X*	*X*	*X*		*X*	*X*					
C. cinnamomea	*X*	*X*		*X*			*X*				
Phellinus gilvus	*X*	*X*	*X*	*X*	*X*	*X*					
Trametes hirta	*X*		*X*		*X*						
Trametes variegata	*X*		*X*		*X*	*X*					
Polyporus albiceps	*X*		*X*		*X*						
P. anisoporus	*X*		*X*		*X*						
P. radicatus	*X*		*X*		*X*						
Lentinus arcularius	*X*	*X*	*X*		*X*	*X*	*X*				
Lentinus tricholoma	*X*		*X*		*X*	*X*					
Trametes biforme	*X*		*X*		*X*						
Trametes versicolor	*X*	*X*	*X*		*X*	*X*					*M*
Cotilidia diaphana	*X*	*X*		*X*		*X*					
Clavariadelphus pistillaris		*X*		*X*							
C. truncatus		*X*		*X*							
Ramaria stricta	*X*	*X*		*X*							
Ramaria botrytis		*X*		*X*							*Y*
Gleophyllum sepiarium	*X*	*X*		*X*							
Panellus stipticus	*X*	*X*	*X*	*X*	*X*	*X*	*X*				
Hygrocybe conica	*X*	*X*		*X*							
Gliophorus psittacinus		*X*		*X*							
Hygrophorus russula		*X*		*X*							
H. chrysodon	*X*	*X*	*X*								

H. hypothejus		*X*		*X*							
Hygrophoropsis aurantiaca	*X*		*X*			*X*					*T*
Hymenopellis radicata	*X*		*X*		*X*						
Xerula pudens	*X*		*X*		*X*	*X*					
Armillaria mellea	*X*	*X*	*X*	*X*	*X*	*X*	*X*				*Y*
Desarmillaria tabescens	*X*	*X*	*X*	*X*	*X*						
Infundibulicybe gibba	*X*	*X*	*X*	*X*			*X*				
Rhodocollybia butyracea	*X*	*X*	*X*	*X*			*X*				
Gymnopus dryophylus	*X*		*X*		*X*		*X*				
Gymnopus androsaceus	*X*	*X*		*X*	*X*	*X*	*X*				
Marasmiellus anthocephallus	*X*	*X*	*X*								
Inocybe geophylla	*X*	*X*	*X*	*X*			*X*				
Onnia tomentosa	*X*		*X*								
Laccaria laccata	*X*	*X*	*X*	*X*	*X*		*X*				*Y*
L. bicolor	*X*	*X*	*X*	*X*							
L. proxima	*X*	*X*	*X*	*X*							
L. amethystina	*X*	*X*	*X*	*X*			*X*				
Neolentinus lepideus	*X*	*X*	*X*	*X*							*M*
Lentinellus ursinus			*X*								
Lepista nuda	*X*		*X*				*X*				*Y*
Leucopaxillus gentianeus	*X*	*X*		*X*							
Lentinus crinitus	*X*		*X*		*X*	*X*			*X*		
P. conchatus			*X*		*X*	*X*					
Pleurotus djamour	*X*		*X*		*X*						*Y*
P. dryinus	*X*		*X*								*Y*
Hohenbuehelia petaloides					*X*	*X*					
Xeromphalina tenuipes	*X*	*X*		*X*			*X*				

(continued)

Table 1 (continued)

Species	Vegetation										
	1. BEP	*2. BP*	*3. BE*	*4. BC*	*5. M*	*6. R*	*7. ChE*	*8. P*	*9.MM*	*10.MR*	*Edibility*
Amanita bisporigera	*X*	*X*	*X*								*T*
A. caesarea	*X*	*X*	*X*								*Y*
A. flavoconia	*X*	*X*	*X*		*X*		*X*				
A. pantherina	*X*	*X*	*X*		*X*		*X*				*T*
A. rubescens	*X*	*X*	*X*				*X*				
A. vaginata	*X*	*X*	*X*		*X*	*X*					*Y*
A. verna	*X*	*X*	*X*								*D*
Agaricus silvaticus	*X*	*X*		*X*	*X*						
A. campestris	*X*		*X*		*X*	*X*		*X*	*X*		*Y*
A. xanthodermus	*X*		*X*	*X*	*X*			*X*	*X*		*T*
Hypholoma fasciculare		*X*		*X*							*T*
Cortinarius violaceus	*X*	*X*	*X*		*X*		*X*				*T*
C. evernius	*X*	*X*		*X*							*T*
C. caligatus	*X*	*X*	*X*								*T*
C. caerulescens	*X*	*X*	*X*								*T*
C. glaucopus	*X*		*X*								*T*
C. atrovirens	*X*		*X*								*T*
C. collinitus	*X*		*X*				*X*				*T*
Cortinarius sanguineus	*X*	*X*	*X*	*X*							*T*
Gymnopilus aeruginosus	*X*	*X*	*X*								*T*
Galerin marginata	*X*	*X*		*X*							*T*
Cyclocybe aegerita	*X*		*X*		*X*			*X*			
Flammulina velutipes	*X*		*X*				*X*				*Y*
Crepidotus cinnabarinus	*X*	*X*	*X*								

C. mollis	*X*	*X*	*X*		*X*		*X*				
Coprinus comatus	*X*		*X*			*X*		*X*	*X*	*X*	*Y*
Coprinellus micaceus	*X*	*X*	*X*	*X*	*X*			*X*			
C.Parasola plicatilis	*X*	*X*	*X*	*X*	*X*	*X*	*X*	*X*	*X*	*X*	
Coprinopsis atramentaria	*X*	*X*	*X*								*T, OH*
C.Coprinellus plicatilis	*X*		*X*		*X*		*X*				
Psathyrella candoleana	*X*	*X*	*X*	*X*	*X*	*X*		*X*		*X*	
Protostropharia semiglobata	*X*	*X*	*X*	*X*	*X*	*X*	*X*	*X*	*X*	*X*	
Deconica coprophila	*X*	*X*	*X*	*X*	*X*	*X*	*X*	*X*	*X*	*X*	*T*
Paneolus acuminatus	*X*	*X*		*X*	*X*		*X*				
Volvariella bombycina	*X*		*X*		*X*						
Boletus pinophilus	*X*	*X*		*X*							*Y*
B. luridus	*X*	*X*		*X*	*X*						*T*
Butyriboletus frostii	*X*	*X*		*X*							*Y*
Cyanoboletus pulverulentus	*X*	*X*		*X*							
Hortiboletus rubellus	*X*	*X*	*X*	*X*	*X*	*X*					*Y*
Gyroporus castaneus	*X*	*X*	*X*	*X*	*X*						*Y*
Chroogomphus rutilus		*X*		*X*							
Leccinellum rugosiceps	*X*	*X*		*X*							
L. albellum	*X*	*X*		*X*							
Pulveroboletus raveneli	*X*	*X*		*X*							
Phylloporus rhodoxanthus	*X*	*X*	*X*								
Porphyrellus fumosipes	*X*	*X*		*X*							
T. plumbeoviolaceus	*X*	*X*	*X*								
T. tabacinus	*X*	*X*		*X*							
Xerocomus illudens	*X*	*X*		*X*							
Boletellus pseudochrysenteroides	*X*	*X*		*X*							

(continued)

Table 1 (continued)

Species	Vegetation										
	1. BEP	*2. BP*	*3. BE*	*4. BC*	*5. M*	*6. R*	*7. ChE*	*8. P*	*9.MM*	*10.MR*	*Edibility*
Aureoboletus russellii	*X*	*X*									*Y*
Strobilomyces confusus	*X*	*X*	*X*								
S. strobilaceus	*X*	*X*	*X*		*X*						*Y*
Suillus tomentosus	*X*	*X*		*X*	*X*						
S. brevipes	*X*	*X*		*X*							*Y*
S. pseudobrevipes	*X*	*X*		*X*							*Y*
S. granulatus	*X*	*X*		*X*	*X*						*Y*
S. luteus	*X*	*X*		*X*							*Y*
Russula brevipes	*X*	*X*	*X*				*X*				*Y*
R. cyanoxantha	*X*	*X*	*X*				*X*				*Y*
R. foetens	*X*	*X*	*X*								
R. virescens	*X*	*X*	*X*		*X*						*Y*
R. mariae	*X*		*X*								
R. emetica	*X*		*X*		*X*		*X*				
R. albonigra							*X*				
R. adusta							*X*				
R. olivacea	*X*	*X*		*X*							
R. lutea	*X*	*X*		*X*			*X*				
Lactarius indigo	*X*	*X*	*X*	*X*							*Y*
L. deliciosus	*X*	*X*		*X*							*Y*
L. zonarius	*X*	*X*	*X*								
L. deterrimus	*X*	*X*	*X*								
L. insulsus	*X*	*X*	*X*								
L. rufus	*X*		*X*		*X*						

L. scrobiculatus	X	X		X			X				
L. subdulcis	X	X		X							
Lactifluus volemus	X	X		X			X				Y
L. uvidus	X	X		X							
Omphalotus olearius	X		X								T
Tapinella panuoides	X		X			X					T
T. atrotomentosus	X	X	X	X		X					T
Scleroderma aereolatum	X		X				X				T
S. verrucosum	X	X		X			X				T
S. cepa	X	X	X		X						T
Pisolithus arhizus	X	X	X		X						
Tulostoma albicans					X				X		
Calvatia craniiformis	X	X		X			X	X			M
C. cyathiformis	X	X		X				X			Y, M
Geastrum pectinatum	X	X		X							
G. triplex	X	X		X			X				
G. minimum	X	X		X							
Calvatia candida	X	X		X				X			
Onygena equina	X	X		X				X			
L. perlatum	X	X		X	X	X	X				Y
L. pulcherrimum	X	X	X				X				
L. pyriforme	X	X	X	X							Y
L. umbrinum	X	X	X				X				
Crucibulum laeve	X	X	X	X	X	X			X		
Cyathus olla	X	X	X	X	X						
C. striatus	X	X	X	X	X						
C. stercoreus	X	X	X	X	X	X		X	X	X	

(continued)

Table 1 (continued)

Species	Vegetation										
	1. BEP	*2. BP*	*3. BE*	*4. BC*	*5. M*	*6. R*	*7. ChE*	*8. P*	*9.MM*	*10.MR*	*Edibility*
Phallus hadriani	*X*	*X*	*X*	*X*	*X*						
Lysurus periphragmoides	*X*		*X*		*X*						
Podaxis pistillaris					*X*			*X*	*X*	*X*	
Battarrea phalloides					*X*				*X*		
Phellorinia inquinans					*X*			*X*	*X*	*X*	
Montagnea arenaria					*X*				*X*	*X*	
Rhizopogon zelleri	*X*	*X*	*X*	*X*							
R. roseolus	*X*	*X*	*X*	*X*							
Hymenogaster tener	*X*	*X*	*X*	*X*							
Octavianina asterosperma	*X*	*X*	*X*	*X*							
↓215	*190*	*145*	*144*	*101*	*93*	*50*	*48*	*16*	*14*	*9*	*810 spp*

Vegetation types: 1. Mixed oak-pine forests (BEP) have 190 species followed by 2. pine forests (BP) with 145, 3. oak forest (BE) with 144, 4. coniferous forest (BC) with 101, 5. Scrub (M) "sensu lato" with 93, 6. riparian forest (R) with 50, 7. Oak chaparral (ChE) with 48, 8. Grasslands (P) with 16, 9. Microphylic desert scrub (MM) with 14 and 10. Rosetophilous desert scrub (MR) 9. M = medicinal; T: Poisonous; TOH: Poisonous with alcohol; D = Deadly; Edible = Y

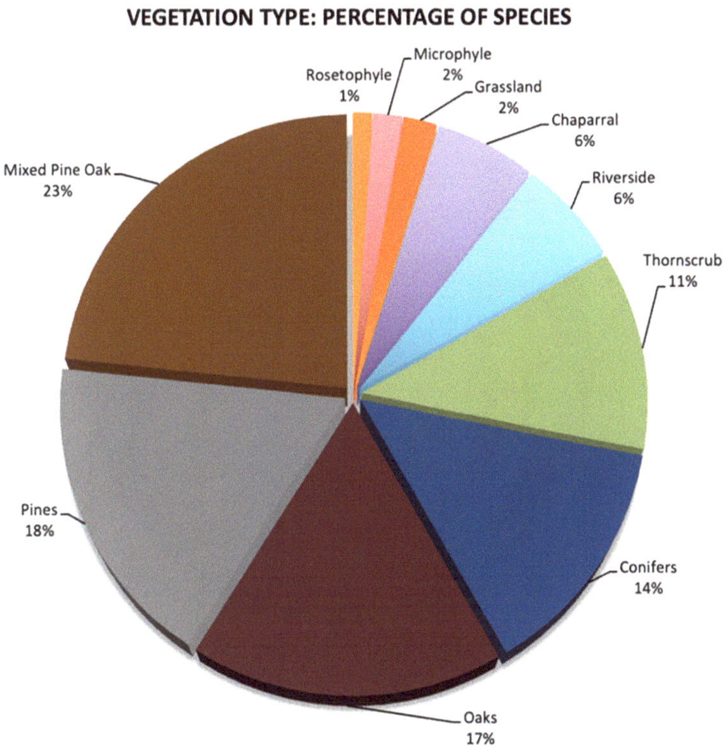

Fig. 2 Distribution of species by type of vegetation

4.1 Associated Hosts and Growth Habit

Many of the fungal species are associated with the floristic component either as parasites, pathogens, saprotrophic or mycorrhizas. Those that are associated with pines and oaks for their forest and economic interest stand out. However, there are many other associations with plant species from the other types of vegetation in the park. Of the species studied, 95 form mycorrhizae with different hosts and some of them are hypogeous with a forest interest (Trappe et al. 2009), 106 are saprotrophic and 14 are parasites of different plants or another fungus as in the case of *Hypomyces lactifluroum* that parasitizes *Russula brevipes* (See Fig. 5).

4.2 Ecological, Economic, Medicinal, Other Uses

Many of the species of fungi of the Cumbres de Monterrey National Park are very important for the type of symbiotic association they form with plant species, some of the species have economic potential since in other countries they are sold in popular

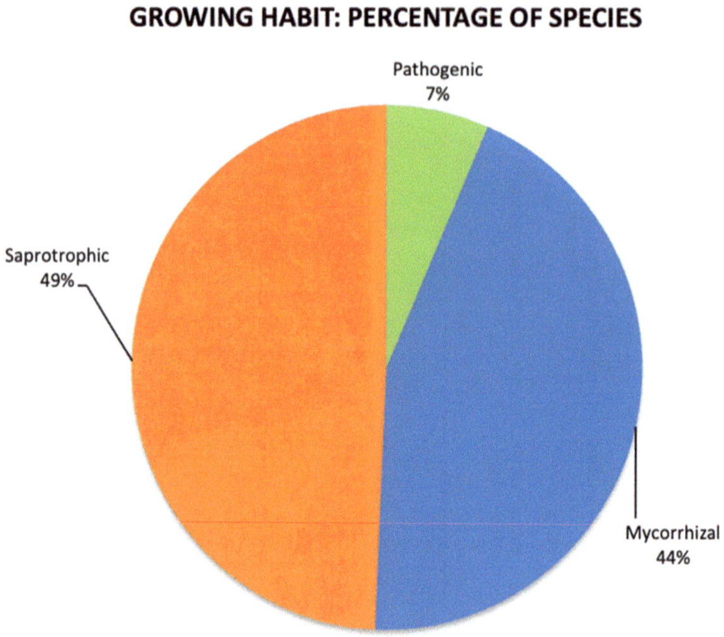

Fig. 3 Percentage of species by type of nutrition (habit) in vegetation types

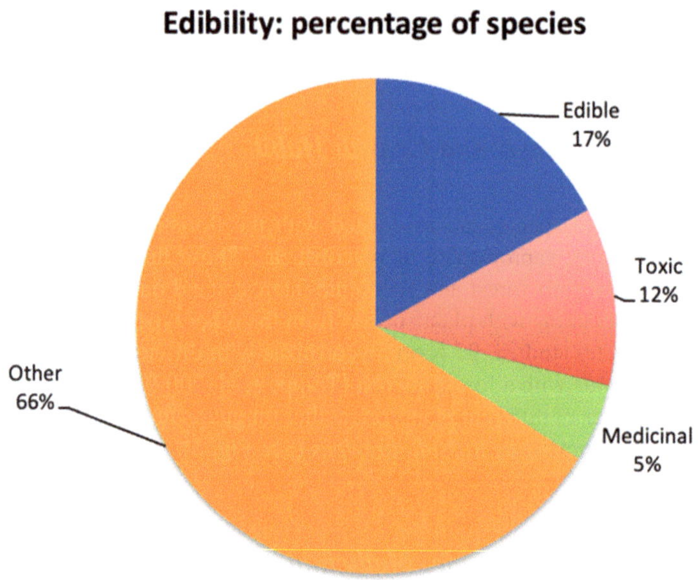

Fig. 4 Edibility of the species and other potential uses

Fig. 5 *Gyromitra esculenta (1), Peziza hemisphaerica (2); Dacryopinax spathularia (3);Trametes versicolor (4); Hexagonia hirta (5); Colticia perennis (6)*

markets such as the fungus *Cantharellus cibarius*, in relation to species with medicinal potential until now they are reported to *Ganoderma applanatum*, *G. lucidum*, *Calvatia cyathiformis*, and *Neolentinus lepideus* among others (Rofuli et al. 2005; Garza et al. 2006; González and Garza 2009). Nothing is known so far about the possible uses that people in rural areas give to fungi in northeastern Mexico. In this study it was determined that 37 species are good quality edible marked as (Yes) in Table 1, 11 have been reported as medicinal in different countries of the world (Med), 25 are reported as poisonous (Come), 1 of them can be toxic when mixed with alcohol (OH) and 1 is deadly (Mort.) (See Fig. 6). There are many

Fig. 6 *Gymnopus butyracea (1); Boletus bicolor (2); Helvella lacunosa (3); Psathyrella sp, (4); Lepista sordida (5); Hohenbuehelia petaloides (6)*

other species that are not poisonous, but they are not considered edible because they are very small, lack good taste and grow solitary and not in groups, also other species are not edible since they have a hard consistency like wood. Some species with these characteristics are used in countries such as Russia, China, Taiwan, and Japan by dry grinding them to then prepare a tea with good flavor and properties to improve the immune system. Finally, some of the species determined from the Cumbres National Park are included in NOM-059-SEMARNAT-2001-2008 so it is necessary to know more about their precise distribution.

5 Discussion

The forests of Cumbres National Park have not been properly managed due to the lack of attention they have had for many years. Currently the forests of the park have suffered strong pressures due to different factors such as the forest fires that occurred in the spring of this year 2021. These fires Caused have significantly affected more than 20 thousand hectares and have left a very unfavorable situation not only in the vegetation that is what is generally thought. Damage includes soil, microscopic animals and large vertebrates, birds, fungi and other living organisms. The trophic chains and the infinity of interactions that all these organisms must feed, grow, reproduce and live are lost. This is how both the forest fires that are increasingly frequent and devastating along with excessive urban growth, clandestine logging, motorized visitors, weekend roast meat visitors, as well as the damage caused by organisms of nature that manifest themselves in the face of ecological imbalance as is the case of the barking insects of some species of pines. All the above is progressively leading to the fragmentation of forest ecosystems, as well as land-use change and a general deterioration of biodiversity. The recreational and extreme tourist activity (motocross, ATVs, mountain biking, rappelling etc.) within the park has increased exponentially in the last 40 years and many of the people who practice these activities do not know nature and are unaware of the ecological damage that their activities promote in the flora, fauna, and fungi of these forest ecosystems. All unnatural changes cause damage that together contribute to the change in the regional climate and to give an example of the results of changes in the forest we can mention the case of bears that come down from the Sierra to look for food and water and have come to be seen in housing complexes and have become a threat to themselves and to the people who live near where they come down from the Sierra. The current forest environment is extremely altered and the effects of the fires of the beginning of this year have already been observed since the first intense rains of the summer dragged in some places, a good part of the little soil and rocky material that was preventing in some cases, the passage of vehicles on roads and highways. The current management forgives offenders of any kind and allows unrestricted entry to all forest sites for apparently recreational purposes and uncontrolled some of the activities that people perform. Thus, anyone can have access to carry out irresponsible tourism activities that often throw away their garbage, cause damage to nature by cutting trees and shrubs for firewood, cause fires and flee after a joyful time of music and roast meat leaving behind a giant problem for the state that without reason does not attend to the corresponding laws. It is unknown how long ecosystems will be able to endure under this type of uncontrolled and aimless management for the conservation of forests and endemic or threatened species. The risks of forest fires are increasing due to the long dry seasons that occur in recent years and all this has an impact on the health of the population. We can act by visualizing restoration activities of the forest masses in the short, medium, and long term and not lose sight of the fact that forests need us so much we need forests for all the goods and services they provide us. As can be seen from the results of this study, the fungi are diverse and were found in all types of vegetation of the park in collections carried out. in the last 40 years. Many species of fungi have economic potential because they are edible, others have pharmacological potential by producing secondary metabolites with action potential for cancer control in humans (Garza et al. 2006; González and Garza 2009). The knowledge of the species, their distribution and their cultivation are giving guidelines to follow new lines of research with species of the park and in Nuevo León new species of truffles with economic potential have been found such as the *Tuber regiomontanum* (Guevara and Garza 2005; Guevara et al. 2008). Of course, forests can be greatly benefited by the proper management of fungal species, especially mycorrhyzal, but not leaving behind the saprobes that help processes of reincorporation of organic matter in sites disturbed by different human activities. With the current scenario in the face of the loss of thousands of hectares of forest due to fires, the importance of fungi in the restoration processes of forests increases, especially of mycorrhizal fungal species. The inoculation of seedlings in nurseries and greenhouses is a necessity to encourage them to obtain once in the field, a greater contribution of nutrients from the mineral soil. In the rainy season you can carry out the establishment of restoration plots with mycorrhizal plants located on well-managed terraces that prevent soil erosion and this procedure can help these plants grow better. The search for mycorrhizal fungal fruits in non-incendiary forests for use in solutions such as seedling inoculant in the nursery or directly on terraces in the field in annual periods is visualized as a feasible and important alternative in support of the park's reforestation activities.

6 Conclusions

The diversity of fungal species associated with the different types of vegetation of the Cumbres National Park is high and is directly related to its conservation status. Fungi benefit forest growth and interact with many species in nutrient cycles in the food chain. The areas affected by this year's forest fires lost more than 95% of the diversity of fungi and this requires attention given their importance for the forest where they participate, for example, in the decomposition of organic matter, in the nutrition of plants through mycorrhizae. Studies should continue the diversity of fungal species in the park's unaffected vegetation and the effects the fires had on diversity. Knowing the species of fungi helps to be able to locate them in situ during the appropriate season, to know their hosts, to learn to use them both in activities of use in reforestation techniques, as well as in research to generate more information in different areas of science.

Below are some common species in Cumbres National Park (Figs. 7, 8, and 9).

Fig. 7 *Hortiboletus rubellus (1); Lactarius deterrimus (2); Laccaria laccata (3); Gomphus clavatus (4); Boletus luridellus (5); Lycoperdon perlatum (6)*

Fig. 8 *Gymnopus velutipes (1); Aureoboletus russelli (2); Calvatia cyathiformis (3); Auricularia nigricans (4); Turbinellus floccosus (5); Hericium erinaceus (6)*

Fig. 9 *Armillaria ostoyae (1); Coprinopsis atramentaria (2); Laccaria amethystina (3); Pleurotus cornucupiae (4); Amanita afin cochiseana (5); Polyporus alveolaris (6)*

References

Boa E (2004) Wild edible fungi a global overview of their use and importance to people. Non-wood forest products 17. FAO, Rome

García J, Garza F (2001) Conocimiento de los hongos de la Familia Boletaceae de México. Ciencia UANL 4(3):336–343

Garza F (1986) Hongos ectomicorrícicos en el estado de Nuevo León. Rev Mex Mic 2:197–205

Garza F, García J, Castillo J (1985) Macromicetos asociados al bosque de *Quercus rysophylla* en algunas localidades del centro del estado de Nuevo León. Rev Mex Mic 1:423–437

Garza L, Ramirez X, Garza F, Salinas MC, Waksman N, Alcaraz Y, Torres O (2006) Evaluación de la actividad biológica de extractos acuosos de macromicetos del noreste de México. Ciencia UANL 9(2):166–170

González P, Garza L, Salinas MC, Vera L, Garza F, Ramírez X, Torres O (2009) Producción de metabolitos antimicrobianos, alcaloides y taninos por *Suillus lakei in vitro*. Ciencia UANL XII (1):62–70

Guevara G, Bonito G, Cázares E, Rodríguez J, Vilgalys R, Trappe JM (2008) *Tuber regiomontanum*, new species of truffle from México. Rev Mex Mic 26:17–20

Guevara G, Garza F (2005) Estudio de la subunidad mayor del ADN ribosomal nuclear de algunas especies del género *Cantharellus* de México. Rev Mex Mic 26:21–26

Guzmán G (1998) Inventorying the fungi of México. Biodivers Conserv 7:369–384

Marmolejo JG, Valenzuela R, Garza F, García J (2013) Micobiota. In: Historia Natural del Parque Natural Cumbres de Monterrey. UANL-CONANP, pp 127–132

Rofuli NB, Cruz AGV, Medalla AP, Buenavista MTS (2005) Antimicrobial and antagonistic properties of Ganoderma lucidum. Int J Med Mushrooms 7(3):150

Trappe JM, Molina R, Luoma DL, Cázares E, Pilz D, Smith J, Castellano M, Miller SL, Trappe MJ (2009) Diversity, ecology, and conservation of truffle fungi in forests of the Pacific Northwest. USDA, Forest Service, Pacific Northwest Research Station, General Technical Report PNW-GTR-772. 194pp

Diversity of Symbiosis Between Species of Macrofungi and Insects in the Temperate Forest of Iturbide, Nuevo León

Fortunato Garza-Ocañas, Gonzalo Guevara Guerrero, Gerardo Cuellar Rodríguez, and Lourdes Garza Ocañas

Abstract Mexico has a high biodiversity that is attributed to its varied geography (Challenger, Utilización y conservación de los ecosistemas terrestres de México. Pasado, Presente y Futuro, 1998). CONABIO and CONANP are making great efforts to publicize the species of organisms that inhabit Mexico. The state of Nuevo León is located in the Northeast of Mexico, occupies 3.3% of the country's surface with a total of 64,082 km^2 and in it adjoin the physiographic regions of the Great North American Plain, Sierra Madre Oriental and Coastal Plain of the Gulf. There is a great diversity of plants in the types of vegetation that grow in these regions and the temperate forests stand out where there is a great diversity of species of macromycetes. The latter play an important role in the flow of nutrients in forests and produce sporocarps that release volatile compounds that attract insects (Amaringo-Cortegano et al., Biota Amazon 3(3):54–63, 2013; Amat-García et al., Rev. Acad. Colomb. Cienc. 28(107):223–232, 2004). Mycotropphy by insects contribute to the process of spore dispersal and germination in the forest ecosystem (Lawrence, Insects-fungus interactions. Academy Press, New York, 1989). In the present research we studied the diversity of associations of mycotropphy by insects associated to sporocarps of species of macrofungi in the temperate forest in the municipality of Iturbide, Nuevo León. The results show that 3680 specimens of 123 species of macrofungi belonging to the phylum Ascomycota and Basidiomycota were collected. The species identified correspond to 6 orders, 19 families and

F. Garza-Ocañas (✉)
Laboratorio de Micología, Facultad de Ciencias Forestales Universidad Autónoma de Nuevo León, Linares, Nuevo León, Mexico

G. G. Guerrero
Tecnológico Nacional de México. Instituto Tecnológico de Ciudad Victoria, Ciudad Victoria, Tamaulipas, Mexico

G. C. Rodríguez
Laboratorio de Entomología, Facultad de Ciencias Forestales Universidad Autónoma de Nuevo León, Linares, Nuevo León, Mexico

L. G. Ocañas
Departamento de Farmacología y Toxicología, Facultad de Medicina, Universidad Autónoma de Nuevo León, Monterrey, Nuevo León, Mexico

F. Garza-Ocañas (ed.), *Sustainable Management of Natural Resources*, Earth and Environmental Sciences Library, https://doi.org/10.1007/978-3-031-33394-1_12

38 genera. The phylum Basidiomycota has greater presence in the sites with 68 of the 73 species identified belonging to 5 orders, 18 families and 34 genera; and the phylum Ascomycota with only 5 species recorded so far, belonging to 1 order, 1 family and 4 genera. From the collected sporocarps were obtained 947 individuals of insects, most of them I.e., 847 individuals belong to the order Coleoptera, and were followed by Hymenoptera with 12, Hemiptera and Lepidoptera with 3 individuals and finally Thysanoptera with 1. The results highlight that the order Coleoptera is strongly associated with sporocarps of Basidiomycota and Ascomycota both saprotrophic, parasites, pathogens and mycorrhizal.

Keywords Mushrooms · Sporocarps · Insects · Associations

1 Introduction

In the state of Nuevo León there are oak, pine and mixed forests that are associated with a high diversity of species of macromycetes that produce fleshy, woody, or corky sporocarps which release spores into the forest. These also release volatile aromatic compounds that betray their presence and serve as attractants for a great diversity of insect species, and other species of animals (e.g. rodents) (García et al. 2005). Macromycetes fulfill functions of recycling organic matter in the forest and to achieve this they developed symbiosis strategies for their feeding e.g. mycorrhizal, saprotrophic, parasites and pathogens (Hernández Santiago et al. 2020). Mycological studies carried out on the vegetation of the state of Nuevo León in recent years reveal the presence of approximately 1300 species of macromycetes (Garza 1986; Herrera 1994). Forest fires and immoderate logging, as well as housing developments and all-terrain vehicle tourism used in ecologically important forested areas are the main causes of affectation to the state's forests. The study of the ecological relationships that exist in forests such as those that occur between forest tree species with macromycete and insect species helps to know and measure how the forest nutrition cycle works in this type of associations (Birkemore et al. 2018; Fogel and Stewart 1975; Guzmán 1994; Hawksworth 1991). It is well known that temperate forest macromycete species are very relevant to tree growth and development given their global contribution to the contribution of mineral nutrients in nutrition cycles and that many animals including insects are natural dispersers of their spores. The forest tree species require both macromycetes and insects as they are essential for their growth and balanced development.

With this line of research it is intended to generate scientific knowledge for example of where and when to locate the sporocarps (i.e. fruits) of the species of macromycetes that are associated with the species of oaks and pines in the forest and later to know the species of insects (i.e. in larval and adult form) that are associated with them in mycotrophy (i.e. they use them as food), as incubators of eggs or nurseries for feeding the emerging larvae of the eggs. The associations of insect-fungus species in the different types of forests give us an idea of their functionality in the forest ecosystem. It has been known that there is a preference of some species of macromycetes for certain species of trees. Similarly, some taxonomic groups of insects e.g. Coleoptera have preference for certain species of macromycetes e.g. Polyporales while others e.g Dipetros prefer other fleshy taxonomic groups of macromycetes e.g. *Amanita*, *Armillaria*. Thus, not all species of macromycetes are associated with all species of trees in the forest and not all species of insects are associated with all species of macromycetes. There are also preferences of some species of insects for the sporocarps of certain species of macromycetes and this is also linked to the form of nutrition that macromycetes can have, such as: mycorrhizal, saprotrophic and parasites in the temperate forest of the municipality of Iturbide, Nuevo León. The results of this type of symbiosis between different organisms (insects-fungi) linked to forest tree species are of great importance. The present study aims to generate information about the nutritional interactions that exist in the temperate forest and that involve the trees, fungi and insects that compose them in the predominant climatic environment, as well as in the conditions of altitude and soils that characterize it. It is necessary to pay attention to them for decision-making about the sustainable management and conservation of species in Northeastern Mexico.

2 Methodology

2.1 Study Area

The municipality of Iturbide is located at the south of the state of Nuevo León at an altitude of between 700 and 3000 meters in the highest parts. In this municipality is located the Ecological Campus of the Autonomous University of Nuevo León originally founded as "Bosque Escuela" belonging to the Institute of Forestry and Management of Natural Resources now Faculty of Forestry Sciences. The dominant soils are: Leptosol (95.5%), Calcisol (2.8%) and Kastañozem (1.7%). As far as geology is concerned, there are sedimentary rocks: Limestone (49%), limestone-shale (36%), shale (10%), conglomerate (2.8%), gypsum (1%) and sandstone-conglomerate (0.2%). The average temperature is 12–22 °C and precipitation is 500–800 mm, the climate is divided into: Semi-dry semi-warm (54%), Temperate subhumid with rains in summer, lower humidity (34%), Semi-warm subhumid with rains in summer, lower humidity (11%) and Temperate subhumid with rains in summer, medium humidity (1%). The vegetation types of the municipality are Forest (67%), scrub (28%) and grassland (1%). The oak forests, mixed oak and pine and pine forests arethe mostcharacteristic and 4% of the territory of the municipality is intended for agricultural use. In the temperate forest of oak-pine and oak pine stand out *Quercus rysophylla*, *Q. polymorpha*, *Q. laeta*, *Q. laceyi*, Q. canbyi, *Q. cupreata*, *Q. graciliformis* as well as *Pinus teocote* and *P. pseudostrobus*. These forests and hosts are associated with a high diversity of macromycete species that produce sporocarps with different aromas, colors, sizes, consistencies and textures; all of them have different nutrition habits e.g. mycorrhizal, parasites, saprobs, pathogens.

Fig. 1 Location of the sampling sites at the school forest of the U.A.N.L

Collection sites of fungal material and its associated insects were established in the temperate forest (Fig. 1).

2.2 Sampling and Processing of Sporocarps

Sampling was carried out on 3 plots of 10 m × 15 m (150 m^2) in 2 vegetation types: oak-pine forest, pine-oak forest (Hernández Santiago et al. 2020). The collections of the sporocarps of the macromycete species were carried out in the sampling plots during the rainy season since it is when they are produced abundantly. 5 random transects were carried out in each of the plots, and in them the sporocarps were labeled and recorded, and the number of species was counted. All fungal material was placed in waxed paper bags along with their field identification tag. The collected specimens were placed in coolers and taken to the laboratory to select a part for processing as nursery camera traps and others for dehydration and subsequent identification. The fruits that were intended for dehydration were placed along with their label on a stove at 80 degrees Celsius for 8 h (Figs. 2 and 3).

Fig. 2 Temperate oak forest – pine in the study area

2.3 *Camera Traps for Collecting Mature Insects Associated with Sporocarps*

The capture of insects from the fruits of the fungi was carried out both in the field directly and in the laboratory. The sporocarps of the different species of macromycetes were placed in transparent plastic trap jars for the insects that come from the field in them to complete their life cycle to the adult state (Fogel and Stewart 1975). These species were caught and placed in jars labeled with 70% alcohol (Figs. 4 and 5).

2.4 *Identification of Macromycete and Insect Species*

The macromycete species were identified in the Mycology Laboratory of the Faculty of Forestry Sciences and the provisional identification of insect species was carried out in the Entomology Laboratory of the Faculty of Biological Sciences of the Autonomous University of Nuevo León (Pardavé and Terán 1999). Macromycete species were placed in a table following the taxonomic criteria of hierarchy of Kirk et al. (2008) (Figs. 6 and 7) (Tables 1, 2, and 3).

Fig. 3 Insects and mites in macromycete sporocarps in the field and laboratory

2.5 Results

Macromycetes

3680 specimens of 123 species of macromycetes belonging to the phylum Ascomycota and Basidiomycota were collected. The identified species correspond to 6 orders, 19 families and 38 genera.

The phylum Basidiomycota has a greater presence at the sites with 68 of the 73 identified species belonging to 5 orders, 18 families and 34 genera; and the phylum Ascomycota with only 5 species recorded so far, belonging to 1 order, 1 family and 4 genera (Figs. 8, 9, 10, 11, 12, 13, 14, 15, 16, 17, 18, 19, 20, 21, 22, 23, 24, 25, 26, 27, 28, 29, 30, and 31).

2.6 Insects

From the individuals of macromycetes collected, 947 individuals of insects were obtained where the order Coleoptera obtained the highest number with 867 individuals, followed by Hymenoptera with 12, Hemiptera and Lepidoptera with 3 individuals and finally Thysanoptera with 1.

Fig. 4 Camera traps with nursery sporocarps for maturation of larvae of associated insects

Fig. 5 Camera traps with sporocarps

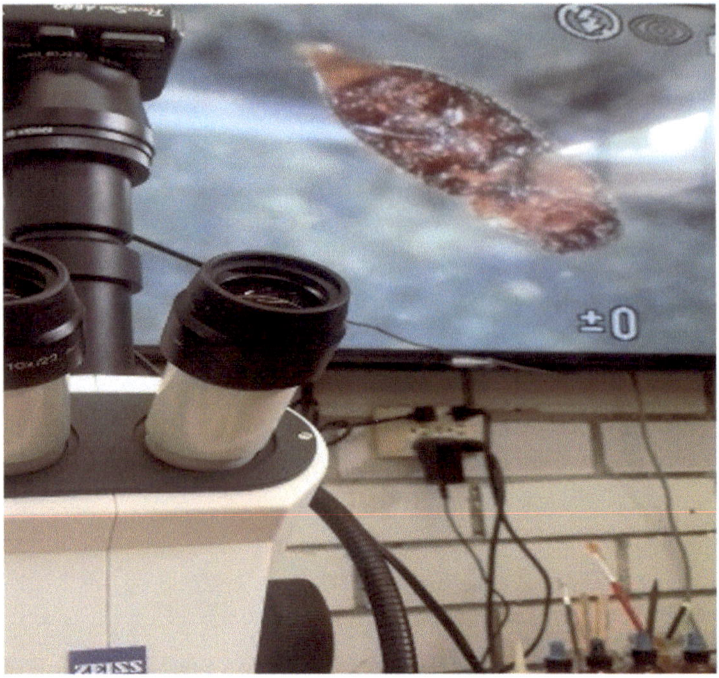

Fig 6 Photography of insects in the laboratory

As already mentioned, the order Coleoptera is the most abundant represented by 867 individuals, where the families Leiodidae, Tenebionidae and Curculionidae are strongly associated with macromycete species. They present a degree of association of mycetobiont type or primary mycophagus since the Coleoptera completed their life cycle in the fruiting body of the fungus using it as an oviposition nursery, creating galleries in the context and chambers for the location of excreta. The order Hymenoptera presented 12 species, where the genus Formicidae was present with the genera *Lasius* and *Tapinoma*. The order Thysanoptera with a single individual, whose identification is still in process, has as genus *Anaphothrips*. The order is Hemiptera and Lepidoptera presented 3 individuals each and so far they are still identified since the species collected for each genus are still unknown (Fig. 32).

2.7 *Identification of Insect Species*

Given the diversity and rarity of the insect species associated with the sporocarps of the macromycetes that had not been previously studied, the species were assigned a provisional key until the determination is made, this is being carried out in the

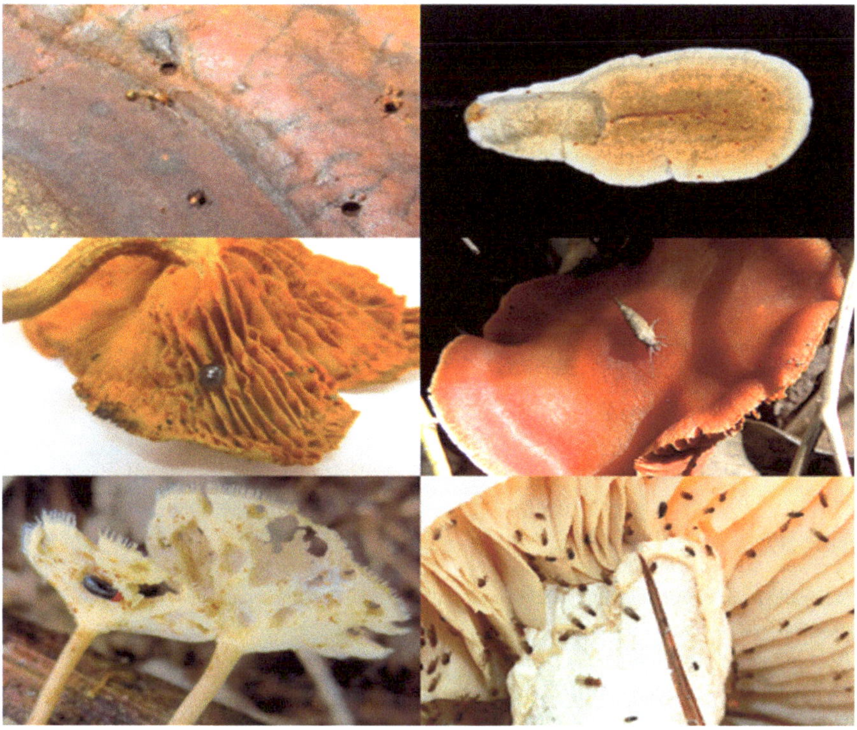

Fig. 7 Examples of insects and mites consuming and using the sporocarps of macromycetes as food or nursery of larvae

Entomology Laboratory of the Faculty of Biological Sciences in charge of Dr. Humberto Quiroz Martínez.

Insect-Fungus Association

The species of macromycetes with the presence of associated insects belong to the phylum Basidiomycota, the species with the presence of a greater number of insects among which the Orders Coleoptera, Lepidoptera, Hymenoptera and Hemiptera stand out are: *Trichaptum biforme, Ganoderma applanatum, Stereum ostrea, Daedalea elegans, Trametes versicolor, T. elegans, T. hirta, T. maxima, Pycnoporus sanguineus, Phellinus gilvus* and *Hexagonia tenuis.*

Insect-Fungus in the Temperate Oak-Pine Forest

Site 2 with vegetation oak forest obtained the largest number of insect individuals with a value of 719 of the 971 total, this represents 74.04%, almost three quarters of the individuals collected so far belong to this collection site, followed by site 1 (TT) with 113 individuals representing 11.63%, at the end is site 3 with vegetation of Oak-Pine Forest which presented the lowest number of insect individuals with only 97 of the total collected, representing only 9.98%.

Table 1 Species of macromycetes, families and growth habit: insects in sporocarps

Species	Family	Habit	Insects
1. Elaphomyces muricatus Vittad	*Elaphomycetaceae*	*M*	*
2. Iodophanus testaceus (Moug.) Korf	*Pezizaceae*		*
3. Hypomyces lactifluorum (Schwein.) Tul. & C. Tul.	*Hypocreaceae*	*M*	*
4. Leucopaxillus gentianeus Fr.	*Tricholomataceae*		*
5. Bovista nigrescens Pers.	*Lycoperdaceae*	*S*	*
6. Calvatia cyathiformis (Bosc) Morgan	*Lycoperdaceae*	*S*	*
7. Calbovista subsculpta Morse ex M.T. Seidl	*Lycoperdaceae*	*S*	*
8. Lycoperdon marginatum Vittad.	*Lycoperdaceae*	*S*	*
9. Bovista pusilla (Batsch) Pers.	*Lycoperdaceae*	*S*	*
10. Calvatia cyathiformis (Guers.) Desv.	*Lycoperdaceae*	*S*	*
11. Panaeolus antillarum (Fr.) Dennis	*Agaricales*	*S*	*
12. Amanita verna (Bull.) Lam.	*Amanitaceae*	*M*	*
13. Amanita bisporigera (G.F. Atk.) E.J. Gilbert	*Amanitaceae*	*M*	*
14. Crepidotus mollis (Schaeff.) Staude	*Crepidotaceae*	*S*	*
15. Inosperma calamistrata (Britzelm.) Lapl.	*Inocybaceae*	*M*	*
16. Pholiota squarrosoides (Peck) Sacc.	*Strophariaceae*	*S*	*
17. Tricholoma caligatum (Viv.) Ricken	*Tricholomataceae*	*M*	*
18. Clitocybe dealbata (Sowerby) P. Kumm.	*Tricholomataceae*	*S*	*
19. Auricularia nigricans (Bull.) Quél.	*Auriculariaceae*	*S*	*
20. Aureoboletus russellii (Frost) G. Wu & Zhu L. Yang	*Boletaceae*	*M*	*
21. Astraeus hygrometricus (Pers.) Morgan	*Diplocystidiaceae*	*M*	*
22. Pisolithus arhizus (Scop.) Rauschert	*Sclerodermataceae*	*M*	*
23. Scleroderma texense Berk.	*Sclerodermataceae*	*M*	*
24. Suillus luteus (L.) Roussel	*Suillaceae*	*M*	*
25. Cantharellus cibarius Fr.	*Hydnaceae*	*M*	*
26. Clavariadelphus truncatus Donk	*Clavariadelphaceae*	*S*	*
27. Ramaria sp.	*Gomphaceae*	*S*	*
28. Trichaptum biforme (Fr.) Ryvarden	*Hymenochaetales*	*P*	*
29. Pseudoinonotus dryadeus (Pers.) T. Wagner & M. Fisch.	*Hymenochaetaceae*	*P*	*
30. Phellinus rimosus (OFrk.) Pilát	*Hymenochaetaceae*	*P*	*
31. Phellinus gilvus (Schwein.) Pat.	*Hymenochaetaceae*	*S*	*
32. Fuscoporia torulosa (Pers.) T. Wagner & M. Fisch.	*Hymenochaetaceae*	*P*	*
33. Antrodia sp.	*Fomitopsidaceae*	*S*	*
34. Laetiporus sulphureus (Bull.) Murrill	*Laetiporaceae*	*P*	*
35. Phaeolus schweinitzii (Fr.) Pat.	*Laetiporaceae*	*P*	*
36. Ganoderma applanatum (Pers.) Pat.	*Polyporaceae*	*P*	*
37. Ganoderma australe (Fr.) Pat.	*Polyporaceae*	*P*	*
38. Ganoderma lucidum (Curtis) P. Karst.	*Polyporaceae*	*P*	*
39. Ganoderma sp.	*Polyporaceae*	*P*	*
40. Abortiporus biennis (Bull.) Singer	*Meruliaceae*	*S*	*
41. Phlebia tremellosa (Schrad.) Nakasone & Burds.	*Meruliaceae*	*S*	*
43. Phlebiopsis papyrina (Mont.) Miettinen & Spirin	*Phanerochaetaceae*	*S*	*

(continued)

Table 1 (continued)

Species	Family	Habit	Insects
44. Trametes versicolor (L.) Lloyd	*Polyporaceae*	*S*	*
45. Cerrena unicolor (Bull.) Murrill	*Cerrenaceae*	*S*	*
46. Lentinus arcularius (Batsch) Zmitr.	*Polyporaceae*	*S*	*
47. Cerrena hydnoides (Sw.) Zmitr.	*Polyporaceae*	*S*	*
48. Perenniporia sp.	*Polyporaceae*	*S*	*
49. Pycnoporus sanguineus (L.) Murrill	*Polyporaceae*	*S*	*
50. Trametes elegans (Spreng.) Fr.	*Polyporaceae*	*S*	*
51. Trametes hirta (P. Beauv.) Zmitr., Wasser & Ezhov	*Polyporaceae*	*S*	*
52. T. maxima (Mont.) A. David & Rajchenb.	*Polyporaceae*	*S*	*
53. Trametes versicolor (L.) Lloyd	*Polyporaceae*	*S*	*
54. T. villosa (Sw.) Kreisel	*Polyporaceae*	*S*	*
55. Bondarzewia Berkeleyi (Fr.) Bondartsev & Singer	*Bondarzewiaceae*	*S*	*
56. Russula cyanoxantha (Schaeff.) Fr.	*Russulaceae*	*M*	*
57. Russula mexicana Burl.	*Russulaceae*	*M*	*
58. Russula sanguinea Fr.	*Russulaceae*	*M*	*
59. Russula emetica (Schaeff.) Pers	*Russulaceae*	*M*	*
60. Stereum gausapatum (Fr.) Fr.	*Stereaceae*	*S*	*
61. Stereum ostrea (Blume & T. Nees) Fr.	*Stereaceae*	*S*	*
62. Stereum hirsutum (Willd.) Pers.	*Stereaceae*	*S*	*

3 Discussion and Conclusions

The results of this study show that there are many nutritional interactions taking place between plant-fungal-insect species in the temperate forest of the municipality of Iturbide. The diversity of fungal species is high and their forms of nutrition (i.e. mycorrhizal, saprobium, parasite or pathogen) or their toxicity do not prevent many insect species from feeding on them or using them as a nursery for their larvae. These results agree with Palacios Vargas and Gómez (1991) who report that the relationships between insects with the fruits of fungi in forests are very common and abundant. The diversity of insect species is also very high and is reported here for the first time in Mexico. The ecological role of interactions between species of fungi and insects is very large because some recycle others but, in their lifecycle, help the growth of forest trees (Challenger 1998; Amaringo-Cortegano et al. 2013; Amat García et al. 2004; Lawrence 1989). For example, mycorrhizal fungi transport mineral nutrients from the soil to plant tissues in their growth and development. The fruits of the fungi release spores by the millions and these are dispersed by the wind, water, and animals among them are insects. The latter propagate the spores of mycorrhizal fungi throughout the forest and they when germinating will form new individuals in new parts of the forest to continue with the activities in seasonal or annual cycles. Interactions between insects and fungi fruits are functionally important in nutrient cycles on the forest floor as well as in the biology of the species involved.

Table 2 Site 1 species of macromycetes in the oak-pine forest and observation of insects in sporocarps

Family	Species	Presence
Agaricaceae	*Agaricus silvicolae-similis Bohus & Locsmándi*	*
	Lepiota clypeolaria (Bull.) P. Kumm.	
	Leucoagaricus rubrotinctus (Peck) Singer	
	Leucocoprinus fragilissimus (Ravenel ex Berk. & M.A. Curtis) Pat.	
Lycoperdaceae	*Lycoperdon echinatum Pers.*	*
Amanitaceae	*Amanita bisporigera (G.F. Atk.) E.-J. Gilbert.*	
	Amanita vaginata (Bull.) Lam.	*
	Amanita verna (Bull.) Lam.	*
Auriscalpiaceae	*Artomyces pyxidatus (Pers.) Jülich*	
	Auriscalpium vulgare Gray	*
Boletaceae	*Hortiboletus rubellus Krombh.*	*
	Aureoboletus russellii (Frost) Murrill, Mimeo	*
	Strobilomyces strobilaceus (Scop.) Berk.	
Hydnaceae	*Cantharellus cibarius Fr.*	*
	Cantharellus sp.	
Callistosporiaceae	*Callistosporium sp. Singer*	
Agaricaceae	*Coprinus comatus (O.F. Müll.) Pers.*	
Psathyrellaceae	*Coprinopsis lagopus (Fr.) Redhead, Vilgalys & Moncalvo*	
	Parasola plicatilis (Curtis) Redhead, Vilgalys & Hopple	
Cortinariaceae	*Cortinarius purpureus (Bull.) Bidaud, Moënne-Locc. & Reumaux.*	*
Cyphellaceae	*Chondrostereum purpureum (Pers.)*	
Dacrymycetaceae	*Dacryopinax spathularia (Schwein.) G.W. Martin*	
Entolomataceae	*Entoloma sinuatum (Bull.) P. Kumm.*	*
Geastraceae	*Geastrum saccatum Fr.*	*
Gloeophyllaceae	*Gloeophyllum sepiarium (Wulfen) P.Karst.*	*
Gomphidiaceae	*Chroogomphus vinicolor (Peck) O.K. Mill.*	*
Helvellaceae	*Helvella crispa Bull.*	*
Hydnangiaceae	*Laccaria laccata (Scop.) Cooke*	*
Hygrophoraceae	*Hygrocybe coccinea (Schaeff.) P. Kumm.*	
Hymenochaetaceae	*Pseudoinonotus dryadeus (Pers.) T. Wagner & M. Fisch.*	*
Crepidotaceae	*Crepidotus mollis (Schaeff.) Staude*	*
Inocybaceae	*Inosperma calamistratum (Fr.) Matheny & Esteve-Rav*	*
Macrocystidiaceae	*Macrocystidia Joss.*	
Omphalotaceae	*Gymnopus androsaceus (L.) Della Magg. & Trassin.*	
Marasmiaceae	*Marasmius maximus Hongo*	
Cyphellaceae	*Gloeostereum incarnatum S. Ito & S. Imai*	*

(continued)

Table 2 (continued)

Family	Species	Presence
Mycenaceae	*Mycena epipterygia (Scop.) Gray*	*
	Xeromphalina cauticinalis (Fr.) Kühner & Maire	*
Omphalotaceae	*Gymnopus dryophilus (Bull.) Murrill*	*
	Gymnopus fusipes (Bull.) Gray	
Polyporaceae	*Trametes versicolor (L.) Lloyd*	
Tapinellaceae	*Tapinella panuoides (Fr.) E.-J. Gilbert*	
Pleurotaceae	*Hohenbuehelia petaloides (Bull.) Schulzer*	*
Psathyrellaceae	*Coprinopsis sp.*	
	Psathyrella sp.	*
Russulaceae	*Lactarius indigo (Schwein.) Fr.*	*
	Lactarius uvidus (Fr.)Fr.	*
	Russula emetica (Schaeff.) Pers.	
Sarcoscyphaceae	*Sarcoscypha occidentalis (Schwein.) Sacc.*	*
Stereaceae	*Stereum complicatum (Fr.)Fr.*	
	Stereum ostrea (Blume & T. Nees) Fr.	
Schizophyllaceae	*Schizophyllum commune Fr.*	*
	Schizophyllum umbrinum OFrk.	
Strophariaceae	*Deconica coprophila (Bull.) P.Karst.*	
Hymenogastraceae	*Gymnopilus aeruginosus (Peck) Singer*	
Strophariaceae	*Protostropharia semiglobata (Batsch) Redhead, Moncalvo & Vilgalys*	*
Sclerodermataceae	*Scleroderma areolatum Ehrenb*	
Suillaceae	*Suillus granulatus (L.) Roussel*	*
Agaricales	*Clitocybe nebularis (Batsch) P. Kumm.*	*
	Infundibulicybe gibba (Pers.) Harmaja	*
Omphalotaceae	*Rhodocollybia butyracea (Bull.) Lennox*	
Hygrophoraceae	*Hygrocybe coccinea (Scchaeff.) P.Kumm*	
Agaricales	*Lepista sordida (Schumach.) Singer*	
	Lepista nuda (Bull.) Cooke	*
Tricholomataceae	*Leucopaxillus albissimus (Peck) Singer*	
	Leucopaxillus gentianeus (Quél.) Kotl	*
Agaricales	*Melanoleuca sp.*	*
Pseudoclitocybaceae	*Pseudoclitocybe sp.*	
Pleurotaceae	*Resupinatus applicatus (Batsch) Gray*	
Tricholomataceae	*Tricholoma ustale(Fr.) P. Kumm.*	
	Tricholoma sulphureum (Bull.) P. Kumm.	
	Tricholoma terreum (Schaeff.) P. Kumm.	*

Table 3 Site 2 Species of macromycetes in which there was observation of insects in sporocarps

Family	Species	Vegetation
Agaricaceae	*Agaricus sp.*	*OF*
Lycoperdaceae	*Lepiota*	*POF, OPF, OF*
	Leucoagaricus rubrotinctus (Peck) Singer	*POF, OF-P OF*
	Leucocoprinus fragilissimus (Ravenel ex Berk. & M.A. Curtis) Pat.	*POF, OF*
	Lycoperdon echinatum Pers.	
Amanitaceae	*Amanita fulva Fr.*	
	Amanita vaginata (Bull.) Lam.	
	Amanita verna (Bull.) Lam.	
	Amanita virosa Bertill	*OF*
Auriscalpiaceae	*Artomyces pyxidatus (Pers.) Jülich*	
Auriscalpiaceae	*Auriscalpium vulgare Gray*	
Boletaceae	*Hortiboletus rubellus (Krombh.) Simonini, Vizzini & Gelardi*	*OF*
	Cyanoboletus sp.	*OPF*
	Aureoboletus russellii (Frost) G. Wu & Zhu L. Yang	
	Strobilomyces strobilaceus (Scop.) Berk.	
Hydnaceae	*Cantharellus cibarius Fr.*	*OF*
	Cantharellus sp.	*OF*
Callistosporiaceae	*Callistosporium sp. Singer*	
Agaricaceae	*Coprinus comatus (O.F. Müll.) Pers.*	
Psathyrellaceae	*Coprinopsis lagopus (Fr.) Redhead, Vilgalys & Moncalvo*	
	Parasola plicatilis (Curtis) Redhead, Vilgalys & Hopple	*OPF*
Cortinariaceae	*Cortinarius purpureus (Bull.) Bidaud, Moënne-Locc. & Reumaux.*	*POF, OPF, OF*
Cyphellaceae	*Chondrostereum purpureum (Pers.)*	
Dacrymycetaceae	*Dacryopinax spathularia (Schwein.) G.W. Martin*	
Entolomataceae	*Entoloma sinuatum (Bull.) P. Kumm.*	*POF, OPF, OF*
Geastraceae	*Geastrum saccatum Fr.*	
Gloeophyllaceae	*Gloeophyllum sepiarium (Wulfen) P.Karst.*	
Gomphidiaceae	*Chroogomphus vinicolor (Peck) O.K. Mill.*	
Helvellaceae	*Helvella crispa Bull.*	
Hydnangiaceae	*Laccaria laccata (Scop.) Cooke*	*OF*
Hygrophoraceae	*Hygrocybe coccinea (Schaeff.) P. Kumm.*	
Hymenochaetaceae	*Inonotus P. Karst.*	
Crepidotaceae	*Crepidotus mollis (Schaeff.) Staude*	*OF*
Inocybaceae	*Inosperma calamistratum (Fr.) Matheny & Esteve-Rav.*	*POF, OPF, OF*
Macrocystidiaceae	*Macrocystidia Joss.*	*OPF*
Omphalotaceae	*Gymnopus androsaceus (L.) Della Magg. & Trassin.*	*OF*

(continued)

Table 3 (continued)

Family	Species	Vegetation
Marasmiaceae	*Marasmius maximus Hongo*	*OF*
	Marasmius oreades (Bolton) Fr.	
Cyphellaceae	*Gloeostereum incarnatum S. Ito & S. Imai*	
Mycenaceae	*Mycena epipterygia (Scop.) Gray*	*POF, OF*
	Xeromphalina cauticinalis (Fr.) Kühner & Maire	
	Xeromphalina Kühner & Maire	
Omphalotaceae	*Gymnopus dryophilus (Bull.) Murrill*	*POF, OPF, OF*
	Gymnopus fusipes (Bull.) Gray	*OPF, OF*
Polyporaceae	*Trametes versicolor (L.) Lloyd*	
Tapinellaceae	*Tapinella panuoides (Fr.) E.-J. Gilbert*	*OF*
Pleurotaceae	*Hohenbuehelia petaloides (Bull.) Schulzer*	*OF*
Psathyrellaceae	*Coprinopsis sp.*	
	Psathyrella sp.	*OPF, OF*
Russulaceae	*Lactarius indigo (Schwein.) Fr.*	
	Lactarius uvidus (Fr.)Fr.	*OF*
	Russula emetica (Schaeff.) Pers.	*OF*
Sarcoscyphaceae	*Sarcoscypha occidentalis.*	
Stereaceae	*Stereum complicatum (Fr.)Fr.*	
	Stereum ostrea (Blume & T. Nees) Fr.	
Schizophyllaceae	*Schizophyllum commune Fr.*	
	Schizophyllum umbrinum Berk.	
Strophariaceae	*Deconica coprophila (Bull.) P.Karst.*	
Hymenogastraceae	*Gymnopilus aeruginosus (Peck) Singer*	
Strophariaceae	*Protostropharia semiglobata (Batsch) Redhead, Moncalvo & Vilgalys*	
Sclerodermataceae	*Scleroderma areolatum Ehrenb*	*OF*
Suillaceae	*Suillus granulatus (L.) Roussel*	*POF*
Agaricales	*Clitocybe nebularis (Batsch) P. Kumm.*	*POF*
	Infundibulicybe gibba (Pers.) Harmaja	
Omphalotaceae	*Rhodocollybia butyracea (Bull.) Lennox*	
Hygrophoraceae	*Hygrocybe coccinea (Scchaeff.) P.Kumm.*	
Agaricales	*Lepista sordida (Schumach.) Singer*	*OPF*
	Lepista nuda (Bull.) Cooke	*OPF, OF*
Tricholomataceae	*Leucopaxillus albissimus (Peck) Singer*	
	Leucopaxillus gentianeus (Quél.) Kotl.	
Agaricales	*Melanoleuca sp.*	*PF-O*
Pseudoclitocybaceae	*Pseudoclitocybe sp.*	*PF-O*
Pleurotaceae	*Resupinatus applicatus (Batsch) Gray*	
Tricholomataceae	*Tricholoma ustale (Fr.) P.Kumm.*	
	Tricholoma sulphureum (Bull.) P. Kumm.	
	Tricholoma terreum (Schaeff.) P. Kumm.	

*OF = Oak Forest (Plot 3), POF = Pine Oak-Forest (Plot 1), OPF = Oak-Pine Forest (Plot 2)

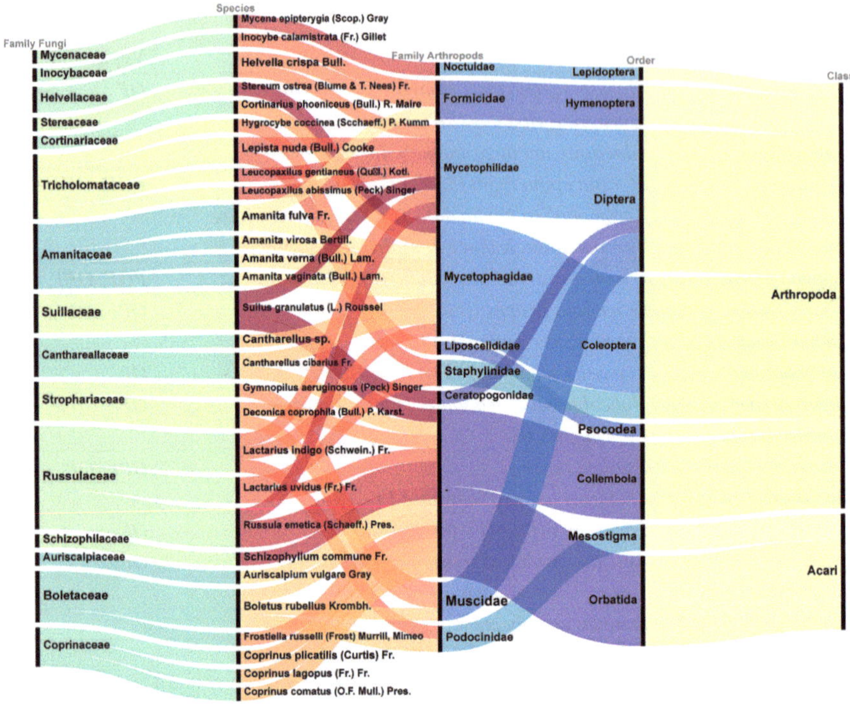

Fig. 8 Interactions between macrofungal and insect species

Fig. 9 *Amanita verna; A. pantherina; Lactarius deliciosus; Hygrocybe punicea; Protostropharia semiglobata; Armillaria mellea; Boletus variipes; Psathyrella sp*

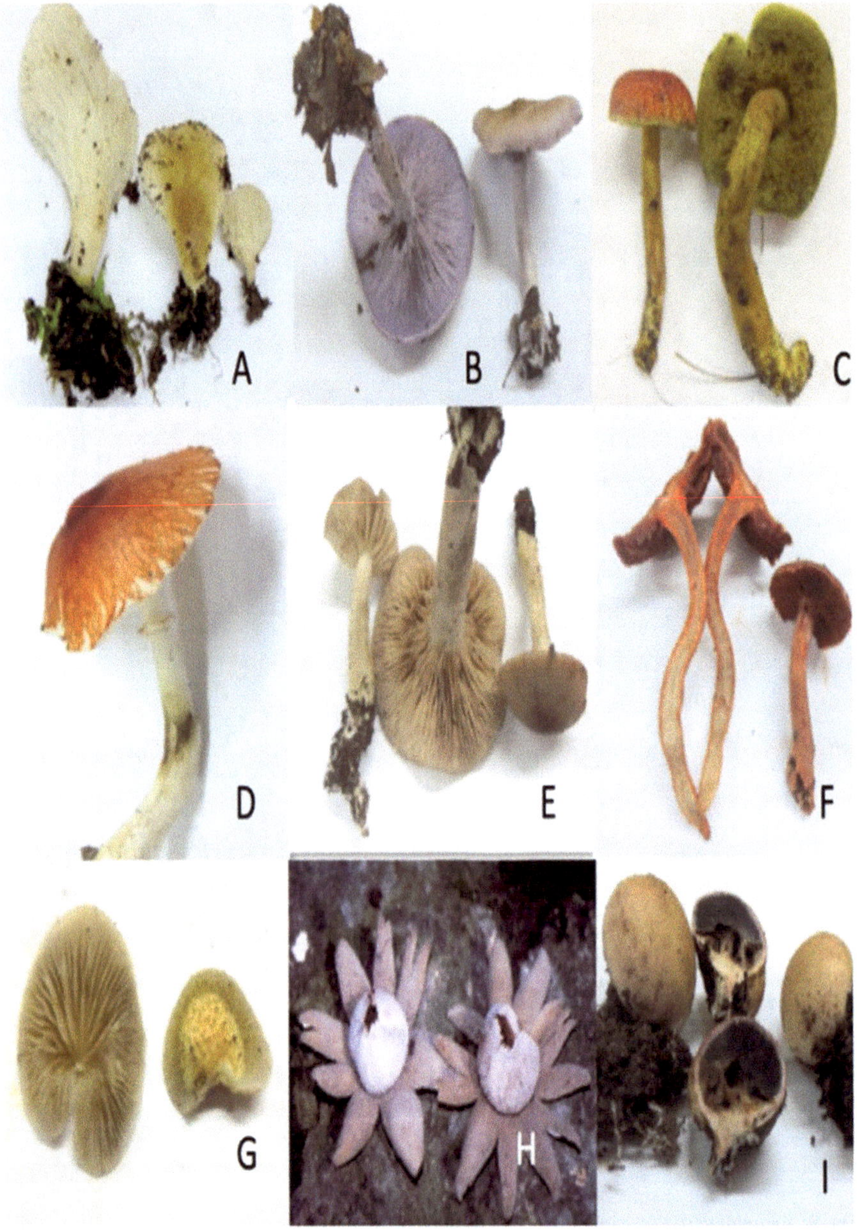

Fig. 10 *Hohenbuehelia petaloides; Lepista sordida; Hortiboletus rubellus; Leucoagaricus rubrotinctus; Entoloma sp., Dermocybe semisanguinea; Crepidotus mollis; Astraeus hygrometricus; Scleroderma areolatum*

Fig. 11 *Cantharellus cibarius; Dermocybe sanguinea; Agaricus silvaticus; Ganoderma resinosum; Byssomerulius incarnatus; Amanita bisporigera; Cortinarius sp; Auriscalpium vulgare; Aureoboletus russelli*

Fig. 12 *Omphalotus subilludens; Helvella crispa; Hohenbuehelia petaloides; Tricholoma ustale; Boletus campestris; Abortiporus biennis; Lactarius indigo; Marasmius androsaceus; Gymnopilus sp*

Fig. 13 *Hygrocybe acutoconica; Inocybe sp; Suillus granulatus; Amanita rubescens; Lepista nuda; Entoloma; Lentinus arcularius; Inocybe rimulosa; Cortinarius violaceus*

Fig. 14 *Stereum ostrea; Lentinus tigrinus; Trametes elegans; Phellinus gilvus; Ganoderma applanatum; Lycoperdon perlatum; Boletus atkinsonii; Coprinellus comatus; Tetrapyrgos nigripes*

Fig. 15 *Gymnopus dryinus; Pycnoporus sanguineus; Trametes versicolor; Lactarius; Leucoagaricus leucothites var. carneifoliu; Rhodogymnopus butiracea; Coprinopsis atramentaria*

Fig. 16 *Tomentella sp.; Boletus bicolor; Boletus variipes; Amanita virosa; Cortinarius sp.; Stereum complicatum*

Fig. 17 *Inocybe calamistrata; Suillellus luridus; Lactarius camphoratus; Cortinarius sp.; Russula sp.; Amanita fulva*

Fig. 18 *Tricholoma ustale; Agaricus silvaticus; Lactarius deterrimus; Amanita afin pantherina; Leucopaxillus gentianeus; Russula emetica*

Fig. 19 *Russula mexicana; Clavulinopsis pixidata; Hygrocybe conica; Marasmius sp; Helvella sp*

Fig. 20 *Pisolithus tinctorius; Lycoperdon echinatum; Russula brevipes; Laccaria laccata; Russula xerampelina; Geastrum triplex*

Fig. 21 *Trametes versicolor; Ganoderma applanatum; Trametes máxima; Dacryopinax spathularia; Inocybe fastigiata; Auricularia nigricans*

Fig. 22 *Humaria hemisphaerica; Helvella sp.; Leotia lubrica; Coprinellus disseminatus; Hydnum repandum; Amanita vaginata*

Fig. 23 *Bovista cyathiformis; Otidea onotica; Gymnopilus aeruginosus; Hypomyces lactifluorum on la brevipes; Helvella crispa; Gloeophyllum striatum*

Fig. 24 *Helvella acetabulum; Artomyces pyxidatus; Geotropa sp.; Cyathus striatus; Helvella sp.; Hydnum repandum var. alba*

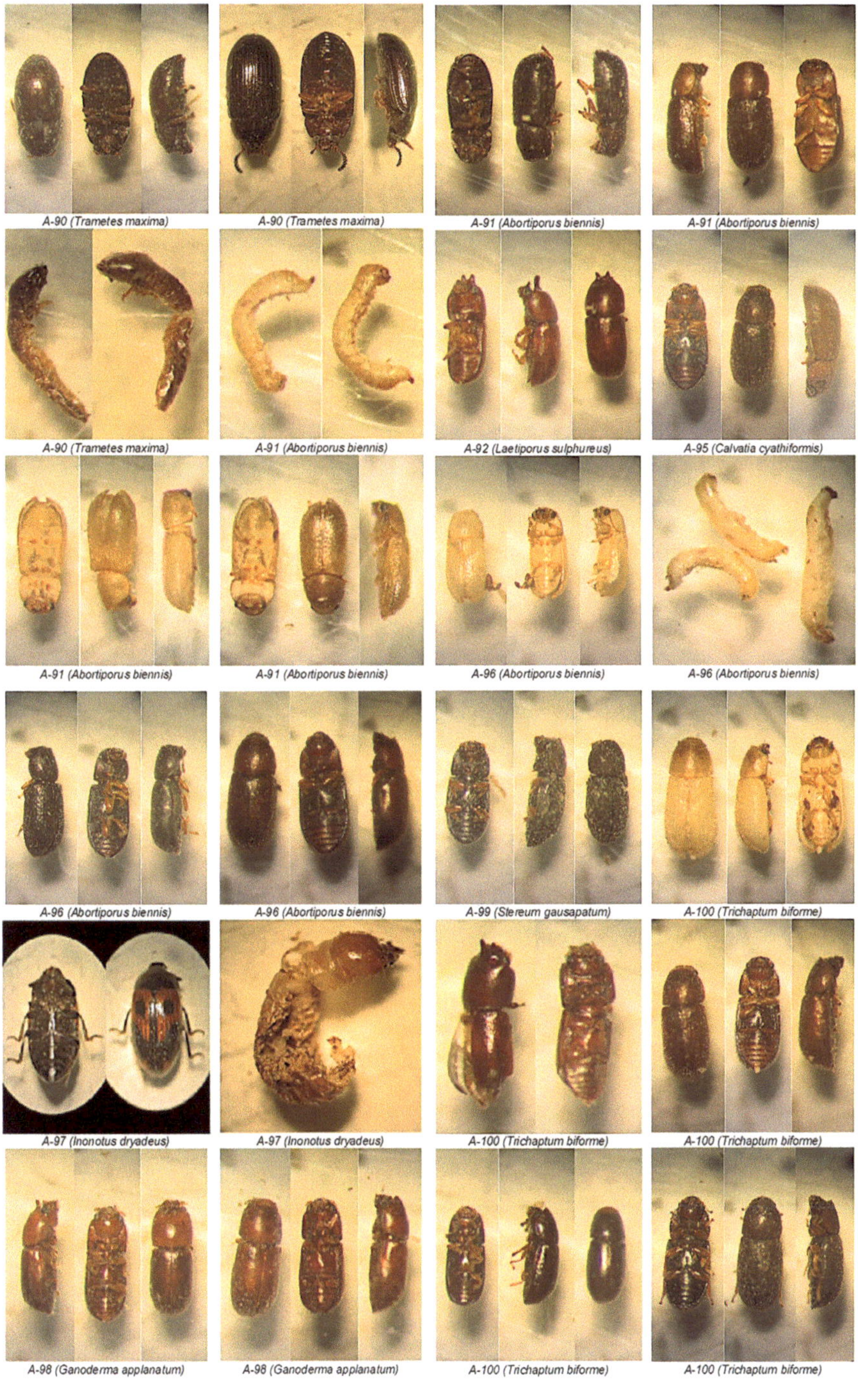

Fig. 25 Insects associated with macromycete sporocarps

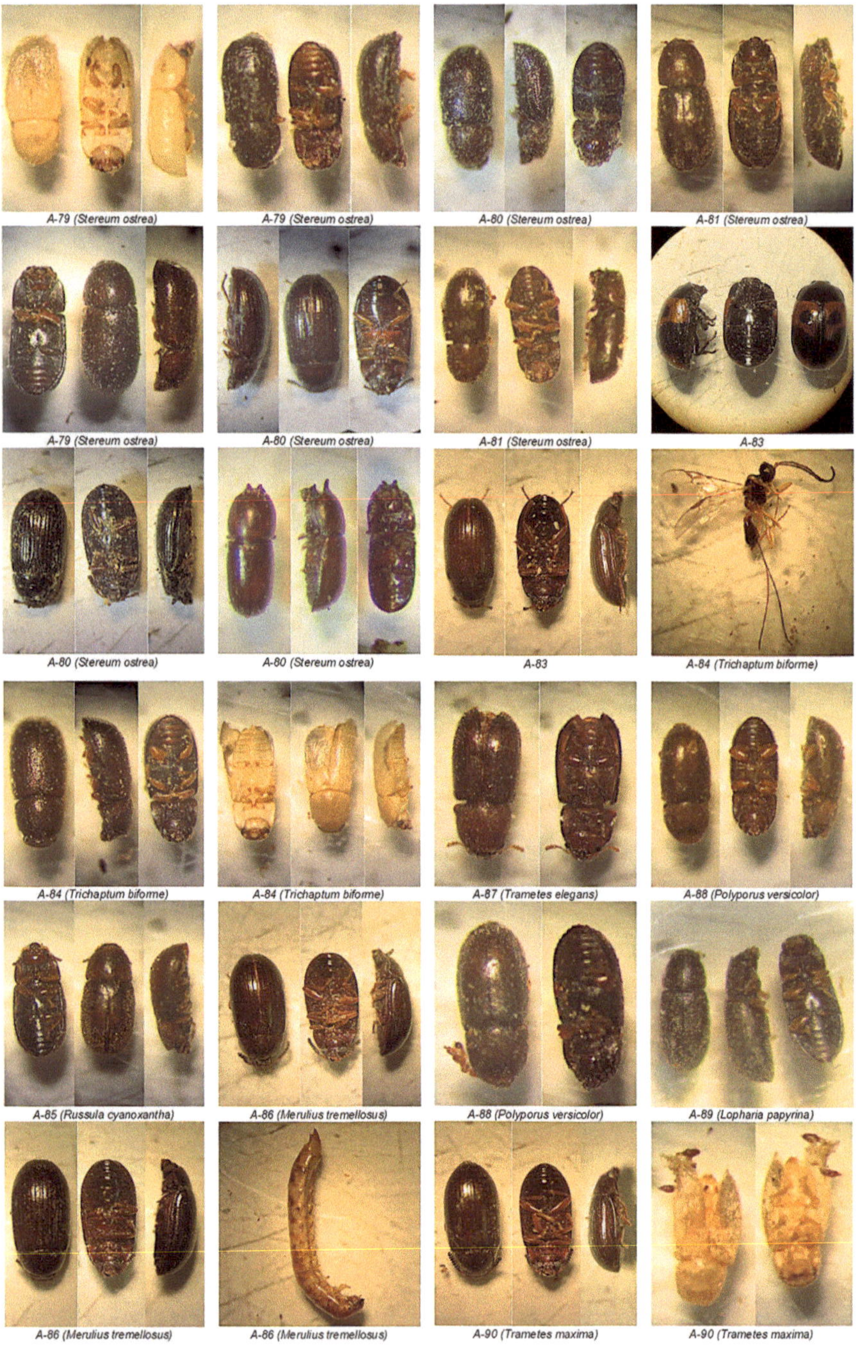

Fig. 26 Insects associated with macromycete sporocarps

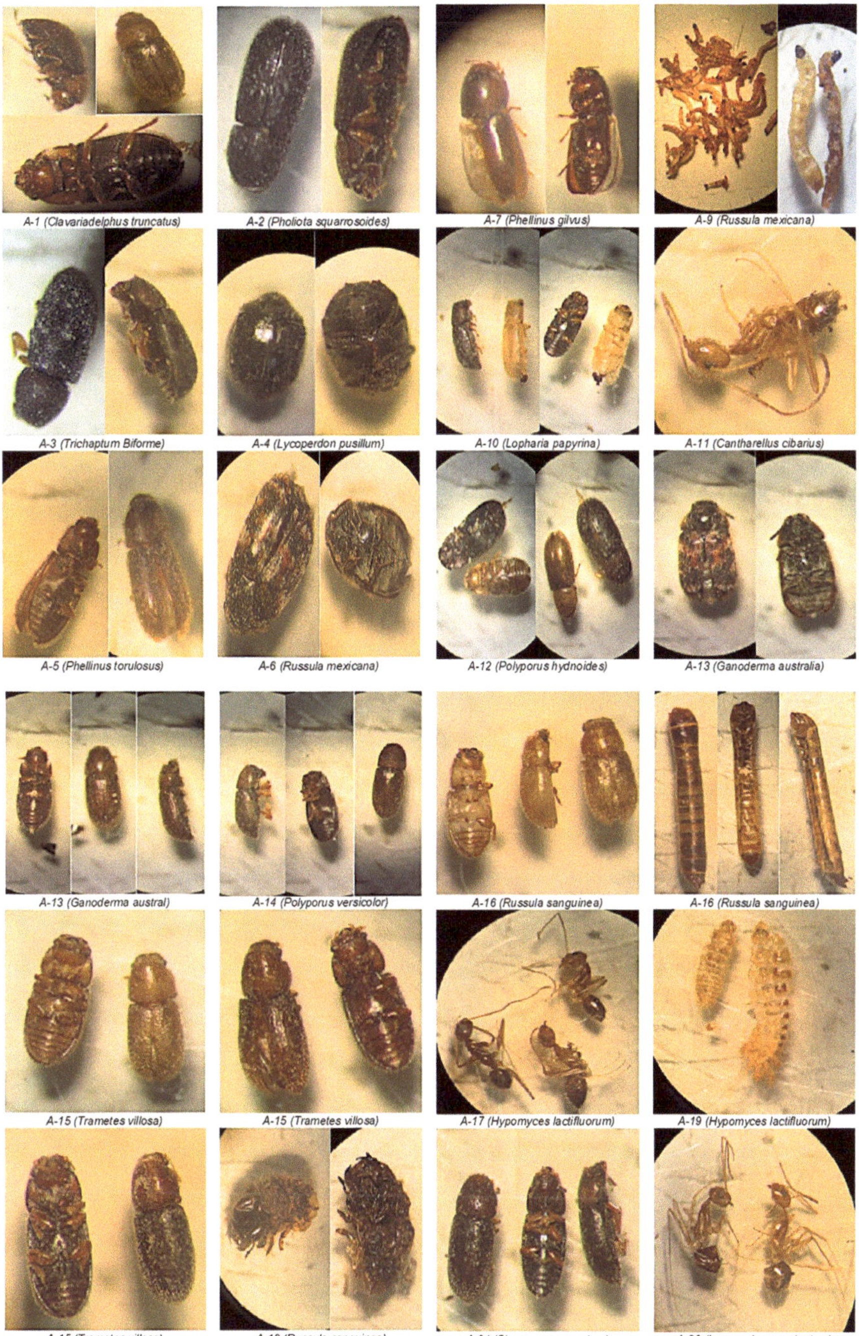

Fig. 27 Insects associated with macromycete sporocarps

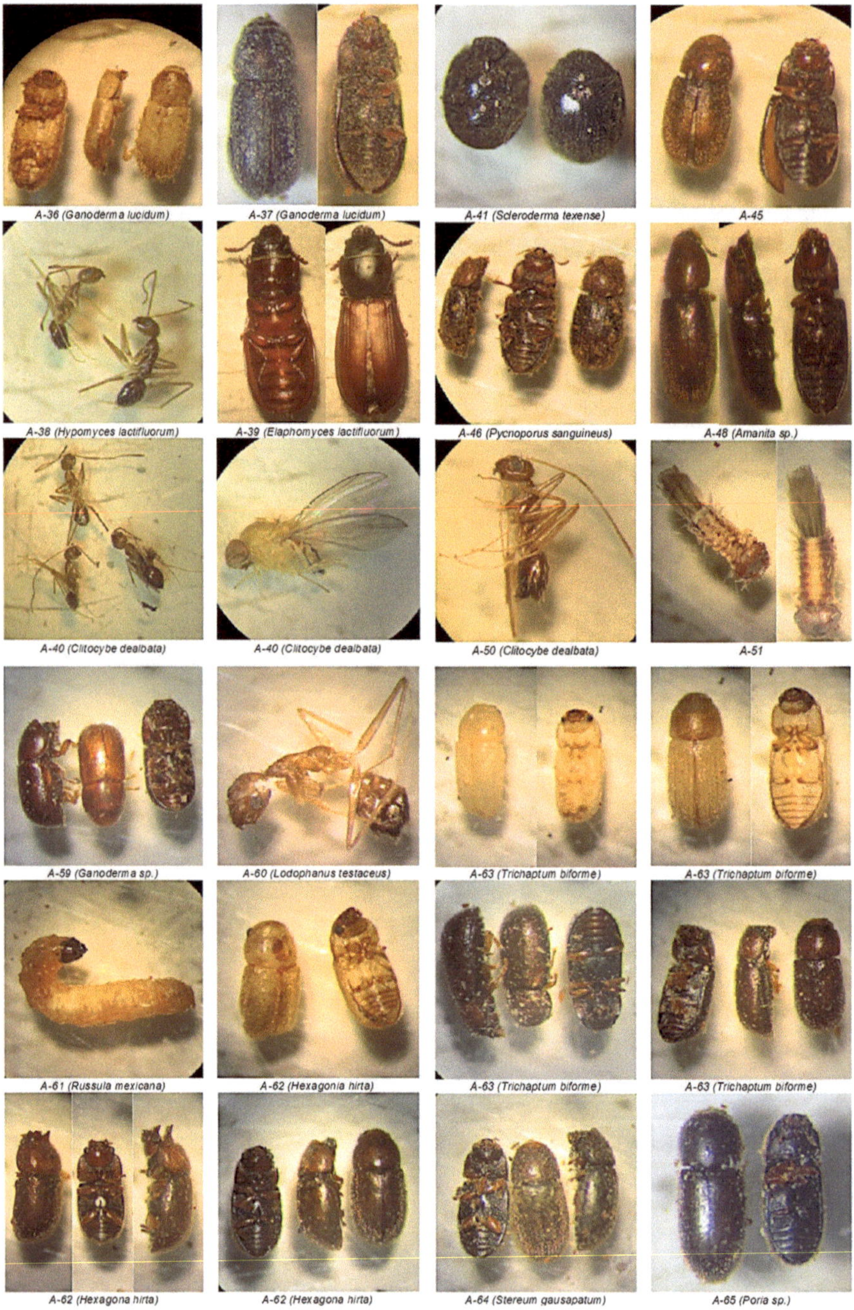

Fig. 28 Insects associated with macromycete sporocarps

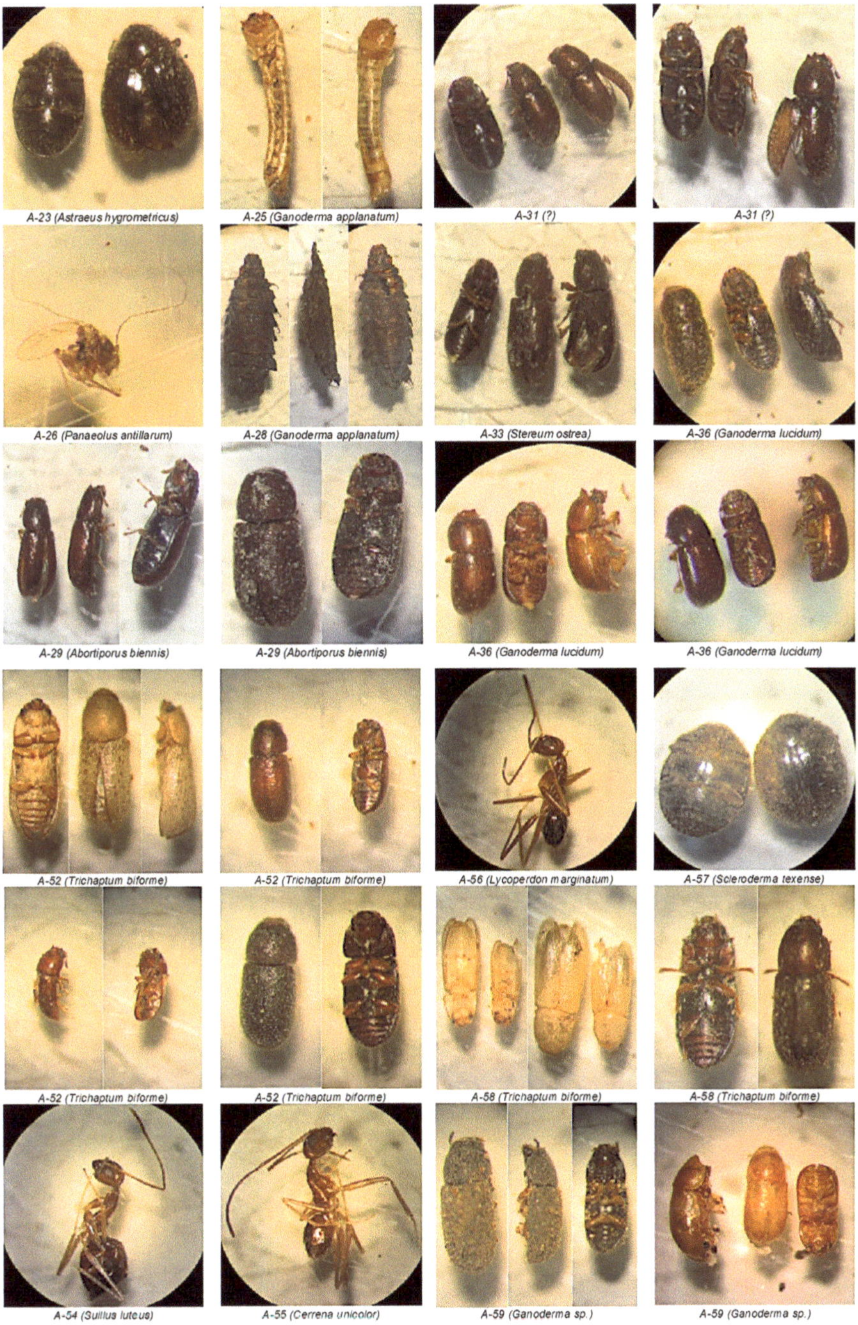

Fig. 29 Insects associated with macromycete sporocarps

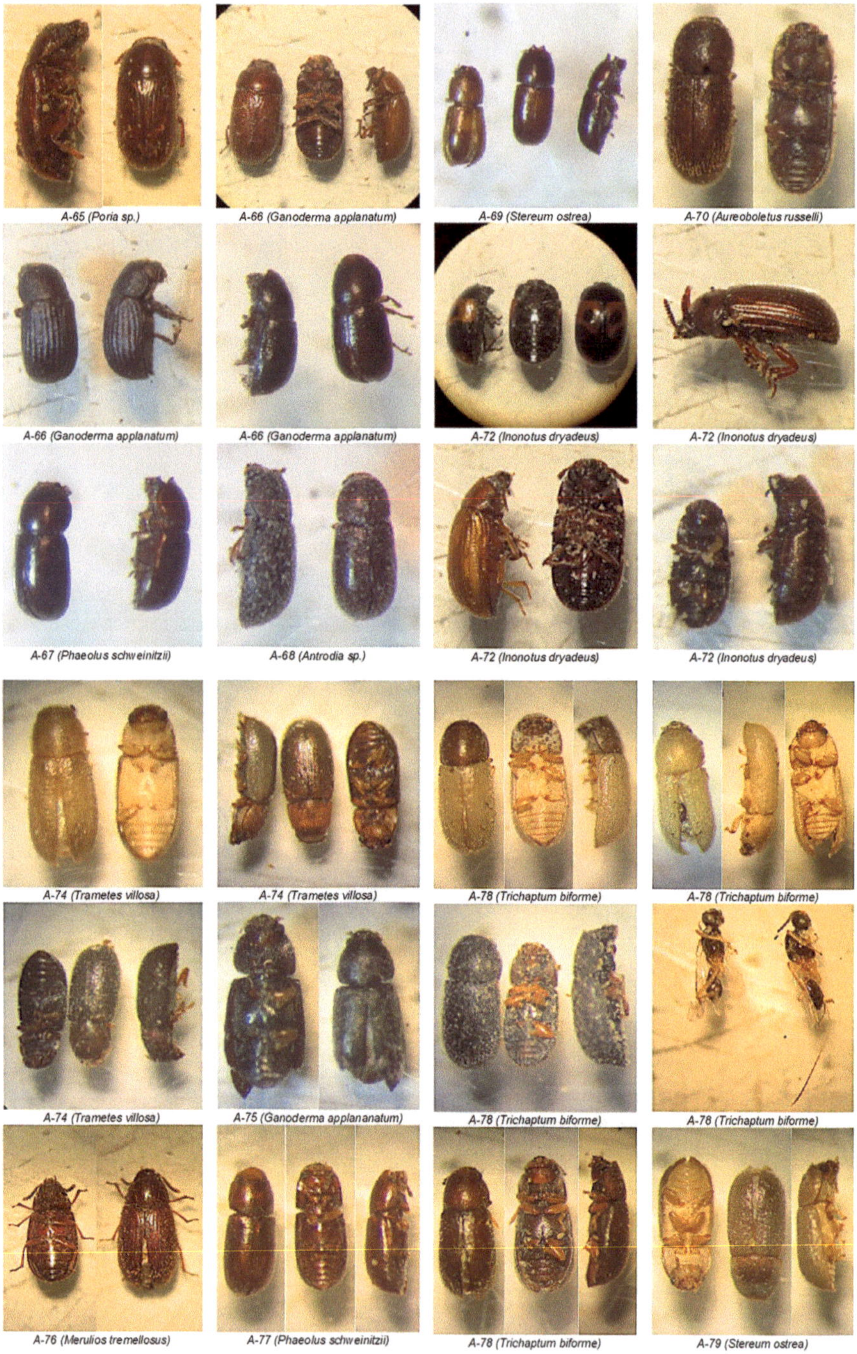

Fig. 30 Insects associated with macromycete sporocarps

Fig. 31 Effects of mycophagia by Coleoptera in a sporocarp of *Ganoderma applanatum*

Fig. 32 *Trichapttum biforme* is the species of fungus that presented a greater number of insect species associated with its sporocarps

References

Amaringo-Cortegano C, Vargas-Islas R, Wellington J, Kazue N (2013) Artrópodos asociados a seis especies de hongos comestibles de ocurrencia natural en Manaus, Amazonas, Brasil. Biota Amazon 3(3):54–63. https://doi.org/10.18561/2179-5746/biotaamazonia.v3n3

Amat-García E, Amat-García G, Henao MLG (2004) Diversidad taxonómica y ecológica de la entomofauna micófaga en un bosque alto andino de la cordillera Oriental de Colombia. Rev Acad Colomb Cienc 28(107):223–232

Birkemoe T, Rannveig MJ, Sverdrup-Thygeson A, Biedermann PHW (2018) Insect-fungus interactions in dead wood systems. Chapter 12 https://www.researchgate.net/publication/325266892 , pp 377–427

Challenger A (1998) Utilización y conservación de los ecosistemas terrestres de México. Pasado, Presente y Futuro. Conabio. 847pp

Fogel R, Stewart B (1975) Ecological studies of hypogeous fungi I. Coleoptera associated with sporocarps. Mycologia 67(4):741–747

Garza F (1986) Hongos ectomicorrícicos en el estado de Nuevo León. Rev Mex Mic 2:197–205

García J, Ramírez Y, Castillo S, Moreno A (2005) Micofagia por roedores en los bosques templados de Tamaulipas. En: Biodiversidad Tamaulipeca. Instituto Tecnológico de Ciudad Victoria 1: 232–236

Guzmán G (1994) Las colecciones de hongos en México y su problemática en la biodiversidad del país. B Soc Bot Méx 55:35–37

Hawksworth DL (1991) The fungal dimension of biodiversity: magnitude, significance and conservation. Mycol Res 95:641–655

Hernández Santiago, F, Díaz Aguilar I, Perez Moreno J, Tovar Salinas JL (2020) Interactions between soil mesofauna and edible ectomycorrhizal fungi. Chapter 14 in Mushrooms humans and nature in a changing world. Perez Moreno et al., editors. Springer Nature Switzerland A.G. 2020. pp 367–405

Herrera T (1994) Perspectivas de la investigación en micología. B Soc Bot Méx 55:39–44

Kirk PM, Cannon PF, Minter DW, Stalpers JA (eds) (2008) Ainsworth & bisby's dictionary of the fungi. CAB International, Wallingford

Lawrence JF (1989) Mycophagy in the coleoptera: feeding strategies and morphological adaptations. In: Wilding N, Collins NM, Hammond P, Webber JF (eds) Insects-fungus interactions. Academy Press, New York, pp 1–23

Palacios Vargas JG, Gómez JA (1991) Los Colémbolos y su relación con los hongos. In Navarrete, J.L. y Quiroz Rocha, G.A. (eds.) Memorias del primer simposio nacional sobre la relación insecto-hongo XXVI Congreso Nacional de Entomología, Veracruz, pp 99–114

Pardavé M, Terán M (1999) Estudio comparativo de dos comunidades de macromicetos en el Área Protegida de Sierra Fría. Revista Investigación y Ciencia 20: 02–10. Zak, J. C. & Willig, M. R. (2004). Fungal biodiversity patterns. In G. M. Mueller, G. F. Bills & M.S. Foster (Eds.), Biodiversity of fungi: Inventory and monitoring methods. (pp 59–75). Londres, Inglaterra: Elsevier Academy

Interactions Between Macrofungi and Insects via Sporocarps in Three Types of Vegetation of the Municipality of Linares, Nuevo León, Mexico

Elisama Rivera-Luna, Fortunato Garza-Ocañas, Humberto Quiroz Martínez, Gerardo Cuellar Rodríguez, José Isidro Uvalle Sauceda, Ricardo Valenzuela Garza, and Gonzalo Guevara Guerrero

Abstract A high diversity of insect-fungus interactions was found in the three vegetation types; these interactions are first reported from northeastern Mexico. Mushroom species grew best in autumn and 102 species were recorded; The Oak Forest (OF) obtained 52 fungi/2206 insect specimens, the Oak-Pine Forest (OPF) obtained 46/1540 and the Tamaulipas Thornscrub (MT) 37/324. The results showed that 4070 insect specimens occurred in 313 individuals of fungi and 35 fungi had 11 species of associated insects. *Daedalea elegans* had 1111 insect specimens and *Hexagonia papyracea*, *Pluteus cervinus* and *Schizophyllum comune* only 1. The beetles had 3809 specimens, Hymenoptera 21, Hemiptera 4, Diptera 3 and Lepidoptera 2. The insect *Colenis* sp, was present in sporocarps of 22 species of fungi and *Lasius niger* in 5 fungi and all the others only in 1. The OF and OPF vegetation types had a similarity of 39.2% (ISS = 0.392) followed by TT with 18.9% (ISS = 0189). The affinity matrix of the Sørensen Coefficient showed that sites 2 and 3 were close with a similarity of 28%, sites 1 and 3 had 10%. The Shannon-Wiener Index showed a high diversity of fungi for all vegetation types; the (OPF) had a diversity of 3.59, the (OF) had 3.57 and the (MT) had 3.07. Regarding the growth habit of the species,

E. Rivera-Luna · F. Garza-Ocañas (✉) · G. C. Rodríguez · J. I. U. Sauceda
Facultad de Ciencias Forestales, Universidad Autónoma de Nuevo León, Linares, Nuevo León, Mexico

H. Q. Martínez
Facultad de Ciencias Biológicas, Universidad Autónoma de Nuevo León. Laboratorio de Entomología, Ciudad Universitaria, San Nicolás de los Garza, Nuevo León, Mexico

R. V. Garza
Laboratorio de Micología, Departamento de Botánica, Escuela Nacional de Ciencias Biológicas, Instituto Politécnico Nacional, Mexico, Mexico

G. G. Guerrero
Tecnológico Nacional de México. Instituto Tecnológico de Ciudad Victoria, Ciudad Victoria, Tamaulipas, Mexico

F. Garza-Ocañas (ed.), *Sustainable Management of Natural Resources*, Earth and Environmental Sciences Library, https://doi.org/10.1007/978-3-031-33394-1_13

67 were Saprotrophic (66%); 25 Mycorrhizal (25%) and 10 parasitic (9%); 49 species were not edible (48%), 27 toxic (26%), 23 edible (23%) and 3 medicinal (3%).

Keywords Macroycetes · Sporocarps · Insects · Seasons · Mycophagy

1 Introduction

Forest macro fungi get their food as saprobes, parasites or mycorrhizal. During the rainy season they produce sporocarps and many species of insects eat or complete their life cycles within these fruits. These plant-fungal-insect interactions are important to the partners involved in these multi-interaction symbioses. Therefore, fungal spores are dispersed by insects to different places where they will produce new mycelium, and insects can obtain food to complete their life cycle and disperse their offspring in the forest. There is no set of individuals, populations or species in ecological isolation, free from interactions with other organisms or populations (Putman 1994), but all organisms are integrated into complex ecological interactions, whose higher-order dynamics (i.e. biodiversity) arise from the interactions between their ecological and biological components (Putman 1994). The biological relationships that can bind organisms within a community are many and varied, many of the links are also extremely subtle or effective (Putman 1994). Many species of fungi are associated with a variety of insects that form different types of symbiosis (Benjamin et al. 2004). In some cases, these associations are obvious; at other times, only exhaustive observations throughout the life cycles of the organisms involved and careful dissection and microscopic examination reveal the presence of an association between them (Benjamin et al. 2004). Talking about insect-fungus relationships is not easy, not only because both groups are considered the most diverse on the planet, but also because the interactions that occur between them can be complex and difficult to contextualize due to the lack of information or in many cases (Delgado and Navarrete-Heredia 2011). Insects use fungal sporocarps as food or as sources of enzymes, this symbiosis allows insects to use nutritional resources and help in the dispersion of fungal spores (Benjamin et al. 2004), some species of fungi such as those of the genus *Phallus,* give off odors to attract flies and other insects that feed on their sporocarps and this is how the symbiosis begins, this type of relationship is called mycophagia (Lawrence 1989; Amat-García et al. 2004). Mycophagia is defined as the consumption of mycelium, fruiting body, spores or any physical structure by an insect (Lawrence 1989); Contrary to popular belief, this dietary pattern is widely disseminated among insects, mainly in Diptera and Coleoptera (Amat 2007). Mycophagia is the main relationship that is established between insects and sporocarps of macromycetes, it is also responsible for shaping the characteristics of the interaction between these two large taxa (Benjamin et al. 2004). According to the degree of association with the fungus, insects can be categorized as: *Mycetobionts:* insects whose association is mandatory, usually depend on the fungus to carry out their life cycle. They use the fungus as a place of refuge and egg position; they are mainly found in the early stages of development of the fungus

Mycetophiles: insects whose dependence on fungi is not absolute; they exhibit some affinity with fungi, but sources of decaying organic matter can also be found in other resources. These types of insects are usually found in intermediate and final stages of fungal development. *Mycetogens:* To this category belong insects that are occasionally found in sporocarps; the relationship is unclear, and they usually use the fungus as a temporary shelter. In this category it is common to find entomophagic insects, parasitoids and hyperparasitoids (Amat 2007). Some Coleoptera complete their life cycle in ephemeral fleshy sporocarps, carry out accelerated larval development of about 3 to 11 days, while those associated with persistent fungi reach within 16 to 72 days; these life stories reflect an adaptation to the nature of the habitat; with these responses they exhibit well-defined patterns according to the use of the resource or the need to face different pressures. Considering preferences Delgado and Navarrete-Heredia (2011) recognize three levels of specificity: *Polyphagia: Insect* species that feed on sporocarps of many different species of macromycetes. Olygophagia*:* A species of insect that feed on sporocarps of a few species, and usually belong to the same genus of macromycetes or to closely related genera. *Monophages:* Insect species that are very selective and specific and only feed on sporocarps of a single species. Most mycetophilous insects are polyphagous, which has been explained by the unpredictable and effective presence of sporocarps, especially agaric ones (Hanski 1989). In contrast, monophagous and oligophagous species are basically found in fungi of the Family Polyporaceae, which produce sporocarps that remain from a few months to several years (Hanski 1989). There are few protocols available for quantitative sampling of fungal-associated insects (Benjamin et al. 2004). The problems involve the unpredictable timing of sporocarp appearance of some species, the microscopic nature of others, the need to cultivate them, and most importantly, the irregular distribution of associated arthropod pruning (Benjamin et al. 2004). The research associated with the diversity of macromycetes for the state of Nuevo Leon is very scarce, and on its associations with other organisms it is almost nil. To determine the value of our forests it is important to have the knowledge of what they harbor. The objective of this research is to determine the diversity and distribution of species of macromycetes and insects that are associated with three types of vegetation, in addition to the degree of association of interaction between them, to observe if there are differences in composition and diversity in each of the sites.

2 Materials and Methods

2.1 Study Area

The municipality of Linares is located southeast of the state of Nuevo León between the geographical coordinates of 25°09' and 24°34' North latitude and 99°07' and 99° 54' longitude Oeste (García-Hernández and Jurado 2008). The study has been carried

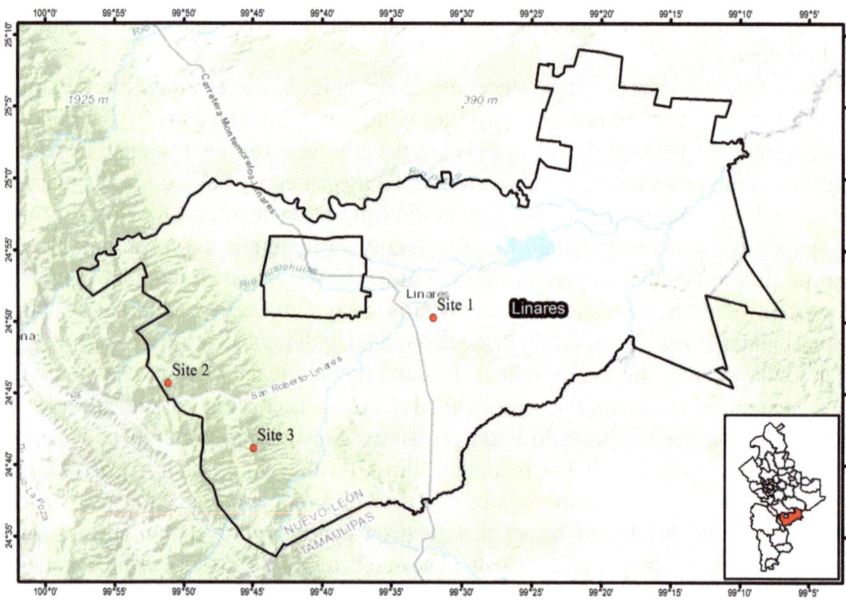

Fig. 1 Location of the study areas

out in three types of vegetation in three sites located in the municipality of Linares. They are the scrub of Tamaulipas (MT) with geographical coordinates 24°47'51"N 99°32'29"W and with an altitude of 379 m. BE Oak Forest with coordinates 24° 41'13.1"N 99°42' 20.5"W and an altitude of 617 m. and Oak Pine Forest (BPE) with coordinates 24°42'50"N 99°47'08"W at an altitude of 1345 m (Fig. 1).

2.2 *Collection of Macromycetes and Associated Insects*

The sampling of the sporocarps was carried out during the four seasons of the year from December 2018 to February 2020 at each of the sites. 3 random transects of 50 meters (3 × 3 × 4) were performed, that is, 3 for the 3 types of vegetation or for the 4 seasons are a total of 36 per year; the transects were randomly located at each of the sites where the topography allowed it. Take photographs before collection are also of greater importance for, in the case of insects, to record their degree of association (e.g. mycetobionts) with macromycete species (Lodge et al. 2004).

The sporocarps of the mushrooms were numbered in the field and collected in plastic bags for transfer to the laboratory in a refrigerator so that they could arrive without decomposing due to the heat. Once in the laboratory, one or more specimens of each species of fungus were placed in plastic bottles of various sizes suitable for the size of each species of mushroom, the top of the bottles was covered with a piece of gauze fixed to the mouth of the bottle with rubber garters to prevent it from falling off. In addition, this fabric allows perspiration. The bottles were labeled with the information corresponding to the fungus species, site, season of the year number of fungi and repetition number and incubated at 25 °C and covered to encourage insects to emerge either as larvae or adults, monitoring of the bottles and collection of insect specimens daily was carried out. The insects collected were placed in small plastic bottles with 70% ethyl alcohol for conservation and subsequent identification.

2.3 Data Analysis

The species of macromycetes and adult insects found in each species of fungus of each collection site were identified, when possible, they were quantified and tabulated hierarchically considering the criteria established by Kirk et al. (2008). Dendrograms of similarity of the species of macromycetes found between the different types of vegetation were made by means of a Cluster Analysis with the MVSP (Multi-Variate Statistical Package) software, which considers the presence/absence of the species in the different sites using the Coefficient of the Sørensen Shannon index determined the diversity and richness of the space of each one. collection site, as well as the entire study.

3 Results

3.1 Diversity of Macromycetes

A total of 313 specimens belonging to 102 species of 71 genera, 12 orders and 35 families belonging to the phylum Ascomycota and Basidiomycota were collected and identified. The phylum Basidiomycota had a greater presence in the vegetation sites with 96 of the 102 species recorded, belonging to 10 orders, 33 families and 66 genera, being the phylum with the greatest weight in all taxonomic levels representing 94.11% of the species collected; it was followed by the phylum Ascomycota with 6 species of 2 orders, 2 families and 5 genera (Fig. 2). The diversity of species was observed in the different types of vegetation and the Oak Forest had the highest number of species with 52 of the 102 collected, it was followed by the Oak Pine Forest with 46 species and at the end the Matorral de Tamaulipas with 37 species (Fig. 3). At all three collection sites, the phylum Basidiomycota was of the highest ecological importance when the largest number of species was recorded. If we consider the exclusive species for each site, that is, those that were recorded only in each of them, the sites (BE) and (BPE) presented the highest number of exclusive species with 24 each, and the site (MT) presented 23 species (Tables 1 and 2).

Regarding the growth habit of the species at sites 67 were saprobes representing 66% of the species, followed by the mycorrhizal with 25 species = 25% and parasites with 10 species = 9%. As for edible, 49 species (48%) were recorded as inedible, 27 toxic (26%), 23 edible (23%) and 3 medicinal (3%).

Fig. 2 Types of vegetation: 1. thornscrub; 2. Oak Forest; 3. Oak-pine forest

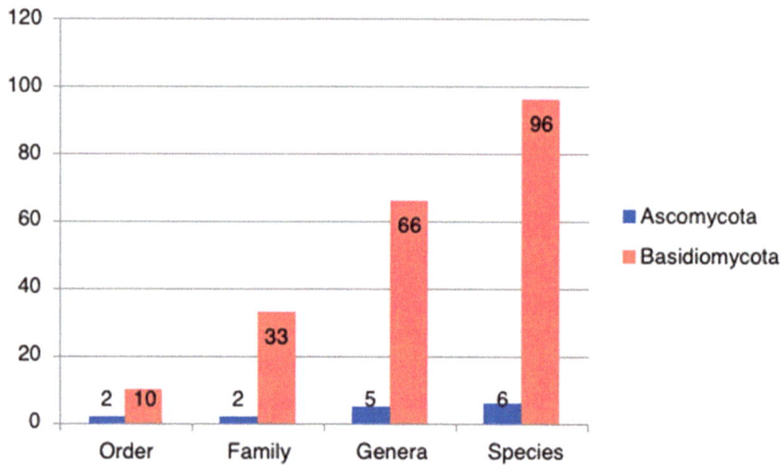

Fig. 3 Taxonomic distribution of the species

3.2 *Insect Diversity*

The results showed that 4070 individuals of insects were obtained from 313 individuals of macromycetes. The order Coleoptera obtained the highest number with 3809 specimens, followed by the order Hymenoptera with 21 individuals, Hemiptera 4, Diptera 3, Lepidoptera 2 individuals, and insects that could not be identified were added in a section called "other" (Fig. 4).

The site (BE) had the highest number of insect individuals associated with macromycetes with 2206 of the totals of 4070, followed by the place (BPE) with 1540 insects y to the final is the place (MT) What Records alone 324 ins cough (Fig. 5).

Table 1 Species of macromycetes exclusive to each type of vegetation: T = Thornscrubs; OF = Oak forest; OPF = Oak-Pine forest

	T	OF	OPF
1	*Daedalea xantha (Fr.) A. Roy & A.B. De*	*Amanita vaginata (Bull.) Lam.*	*Amanita caesarea* (Scop.) Pers.
2	*Campanella* sp.	*Armillaria mellea* (Vahl)P. Kumm.	*Amanita pantherina* (DC.) Krombh.
3	*Crepidotus* sp.	*Astraeus pteridis* (Shear) Zeller	*Amanita verna* (Bull.) Lam.
4	*Daedaleopsis confragosa (Bolton) J.*	*Boletus* sp.	*Antrodia* sp.
	Schröt.		
5	*Fuscoporia ferruginosa* (Schrad.)	*Chondrostereum purpureum* (Persoon)	*Auricularia auricula-judae (Bull.) Quél.*
	Murrill	Pouzar	
6	*Gloeophyllum*	*Cortinarius* sp.2	*Boletus* sp.2
	sepiarium (Wulfen) P.		
	Karst.		
7	*Trametes variegata*	*Entoloma* sp.	*Calocybe cyanea* Singer ex
	(Berk.) Zmitr., Wasser		Redhead & Singer
	& Ezhov		
8	*Pseudofavolus tenuis*	*Exidia glandulosa* (Bull.) Fr.	*Cortinarius* sp.3
	(Fr.) G. Cunn.		
9	*Hypoxylon* sp.	*Gymnopus erythropus*	*Peniophora albobadia*
		(Pers.) Antonín, Halling &	*(Schwein.) Boidin*
		Noordel.	
10	*Annulohypoxylon*	*Hygrocybe* sp.	*Hypomyces chrysospermus*
	truncatum (Starbäck)		Tul. & C. Tul.
	Y.M. Ju, J.D. Rogers		
	& H.M. Hsieh		
11	*Inocutis* sp.	*Inocybe geophylla* (Sowerby) P. Kumm.	*Inocybe* sp.
12	*Lysurus periphragmoides* (Klotzsch) Dring	*Lentinus tigrinus* (Bulliard)Fr.	*Irpex lacteus* (Fr.) Fr.
13	*Neofavolus alveolaris (DC.) Sotome & T. Hatt.*	*Mycena margarita* (Murrill) Murrill	*Lactarius indigo* (Schweinitz) Kuntze
14	*Poria* sp.	*Panaeolina foenisecii* (Pers.) Maire	*Laetiporus sulphureus* (Bull.) Murrill
15	*Psathyrella candolleana* (Fr.) Maire	*Panaeolus antillarum* (Fr.) Dennis	*Panus neostrigosus Drechsler-Santos & Wartchow*
16	*Psathyrella* sp.	*Panellus stipticus* (Bull.) Fr. Karst.	*Zhuliangomyces illinitus (Fr.) Redhead*
17	*Trametes hispida* Bagl.	*Peniophora quercina* (Pers.) Cooke	*Hygrocybe ovina (Bull.) Kühner*

(continued)

Table 1 (continued)

	T	OF	OPF
18	*Tremella lutescens* Pers.	*Phaeocollybia* sp.	*Omphalotus subilludens* (Murrill) H.E. Bigelow
19	*Trametes occidentalis* (Klotzsch)Fr.	*Pluteus longistriatus* (Peck) Peck	*Phyllotopsis nidulans* (Pers.) Singer
20	*Trichaptum biforme* (Fries) Ryvarden	*Pluteus* sp.	*Russula brevipes* Peck
21	*Tulostoma* sp.	*Pluteus* sp.2	*Russula cyanoxantha* (Schaeff.) Fr.
22	*Ustulina deusta* (Hoffm.) Maire	*Pycnoporus* sp.	*Russula mexicana* Burl.
23	*Xylaria hypoxylon* (L.) Grev.	*Russula delica* Fr.	*Russula* sp.
24		*Scleroderma citrinum* Pers.	*Xerocomus subtomentosus* (L.) Quél.

3.3 Insect-Fungus Interactions

Of the 102 species of macromycetes identified, 35 of them presented some type of association with insect species, so *Daedalea elegans recorded* the highest number of insect individuals with 1111 of the 4070 collected, and, on the contrary, the macromycete species that had the lowest number of associated insects were: *Hexagonia papyracea, Pluteus cervinus* and *Schizophyllum commune* with a single insect (Figs. 6 and 7).

As already mentioned, the order Coleoptera is the most abundant represented by 3809 individuals, where the families Leiodidae and Tenebrionidae are strongly associated with the species of macromycetes collected, presenting a degree of association or mycetobiont type or primary mycophagia since the Coleoptera completed thier life cycle in the fruiting body of the fungus uses it as a place for oviposition creating galleries, and small chambers to deposit its waste. The order Hymenoptera had 21 individuals, where the family Formicidae was present with the genus *Lasius,* this genus was recorded forming a type of association Mycetogena. The Order Hemiptera recorded a species of the Family Aradidae that had an association of Micetogenus with the species *Hexagonia tenuis* and *Stereum complicatum.* For the Order Diptera, *Musca domestica* was recorded as micetophillic associated with *Lysurus periphragmoides* collected in the MT. Of the 11 insect species identified, 5 presented a degree of mycetobiont-like association with macromycete species, where we can find *Colenis* sp., *Neomida bicornis, N. haemorrhoidalis, Diaperis* sp. and *D. rufipes*; *Prometopia* sp. and Musca *domestica* had a mychastophile association, a type of Mycetogenic association consisted of *Lasius niger,* Hemiptera of the family Aradidae, and *Ithycerus* sp., of the order Arachnida of the family Ixodidae (Table 3).

Table 2 Diversity of species, growing habit and edibility in each type of vegetation

Family	Species	Habit	Edibility	Sites		
				TT	BE	BPE
Xylariaceae	*Annulohypoxylon thouarsianum* (Lév.) Y.M. Ju, J.D. Rogers & H.M. Hsieh	S	Not edible		x	x
	Annulohypoxylon truncatum (Starbäck) Y.M. Ju, J.D. Rogers & H.M. Hsieh	S	Not edible	x		
	Hypoxylon sp.	S	Not edible	x		
	Ustulina deusta (Hoffm.) Maire	S	Not edible	x		
	Xylaria hypoxylon (L.) Grev.	S	Medicinal	x		
Hypocreaceae	*Hypomyces chrysospermus* Tul. & C. Tul.	P	Not edible		x	
Tremellaceae	*Tremella lutescens* Pers.	S	Edible	x		
Agaricaceae	*Cyathus stercoreus* (Schwein.) By Toni	S	Not edible	x	x	
	Tulostoma sp.	M	Not edible	x		
Amanitaceae	*Amanita caesarea* (Scop.) Pers.	M	Edible			x
	Amanita pantherina (DC.) Krombh.	M	Toxic			x
	Amanita verna (Bull.) Lam.	M	Toxic			x
	Amanita vaginata (Bull.) Lam.	M	Toxic			
	Zhuliangomyces illinitus (Fr.) Redhead	M	Edible			x
Cortinariaceae	*Cortinarius* sp.	M	Toxic		x	x
	Cortinarius sp.2	M	Toxic		x	
	Cortinarius sp.3	M	Toxic			x
	Phaeocollybia sp.	M	Toxic		x	
Cyphellaceae	*Chondrostereum purpureum* (Persoon) Pouzar	S	Not edible	x		
Entolomataceae	*Entoloma* sp.	S	Toxic		x	
Hygrophoraceae	*Hygrocybe* sp.	S	Toxic		x	
	Hygrocybe ovina (Bull.) Kühner	M	Toxic			x
Inocybaceae	*Crepidotus* sp.	S	Not edible	x		
	Inocybe geophylla (Sowerby) P. Kumm.	M	Toxic		x	
	Inocybe sp.	M	Toxic			x
Lycoperdaceae	*Apioperdon pyriforme* (Schaeff.) Vizzini	S	Toxic		x	x

(continued)

Table 2 (continued)

Family	Species	Habit	Edibility	Sites TT	BE	BPE
Lyophyllaceae	*Calocybe cyanea* Singer ex Redhead & Singer	M	Edible			x
Marasmiaceae	*Campanella* sp.	S	Not edible	x		
	Tetrapyrgos nigripes (Fr.) E. Horak	S	Not edible		x	x
Mycenaceae	*Mycena margarita* (Murrill) Murrill	S	Toxic		x	
	Panellus stipticus (Bull.) Fr. Karst.	S	Not edible		x	
Omphalotaceae	*Gymnopus erythropus* (Pers.) Antonín, Halling & Noordel.	S	Toxic		x	
	Omphalotus subilludens (Murrill) H.E. Bigelow	M	Toxic			x
Phyllotopsidace ae	*Phyllotopsis nidulans* (Pers.) Singer	S	Edible			x
Physalacriaceae	*Armillaria mellea* (Vahl)P. Kumm.	P	Edible		x	
	Desarmillaria tabescens (Scop.) R.A. Koch & Aime	S	Edible		x	x
	Oudemansiella melanotricha (Dörfelt)M.M. Moser	S	Edible		x	x
Pleurotaceae	*Resupinatus alboniger* (Pat.) Singer	S	Not edible		x	x
Pluteaceae	*Pluteus cervinus* (Schaeff.) Kumm.	S	Toxic	x		x
	Pluteus longistriatus (Peck) Peck	S	Toxic		x	
	Pluteus sp.	S	Toxic		x	
	Pluteus sp.2	S	Toxic		x	
Psathyrellaceae	*Coprinopsis lagopus (Fr.) Redhead, Vilgalys & Moncalvo*	S	Toxic	x	x	
	Coprinopsis nivea (Pers.) Redhead, Vilgalys & Moncalvo	S	Toxic		x	x
	Panaeolina foenisecii (Pers.) Maire	S	Not edible		x	
	Panaeolus antillarum (Fr.) Dennis	S	Toxic		x	
	Psathyrella candolleana (Fr.) Maire	S	Not edible	x		
	Psathyrella sp.	S	Not edible	x		

(continued)

Table 2 (continued)

Family	Species	Habit	Edibility	Sites		
				TT	BE	BPE
Schizophyllacea e	*Schizophyllum commune* Fr.	S	Edible	x	x	
Strophariaceae	*Deconica coprophila* (Bull.) Fr. Kumm.	S	Toxic	x		x
Auriculariaceae	*Auricularia auricula-judae (Bull.) Quél.*.	S	Edible			x
	Exidia glandulosa (Bull.) Fr.	S	Not edible		x	
Boletaceae	Boletus sp.	M	Not edible		x	
	Boletus sp.2	M	Not edible			x
	Xerocomus subtomentosus (L.) Quél.	M	Edible			x
Sclerodermatac eae	*Scleroderma citrinum* Pers.	S	Toxic		x	
Gloeophyllaceae	*Gloeophyllum sepiarium* (Wulfen) P. Karst.	S	Not edible	x		
Hymenochaetac eae	*Fulvifomes* sp.	P	Not edible			x
	Fuscoporia ferruginosa (Schrad.) Murrill	P	Not edible	x		
	Hydnoporia olivacea (Schwein.) Teixeira	P	Not edible		x	x
	Inocutis sp.	S	Not edible	x		
	Phellinus gilvus (Schwein.) Pat.	P	Not edible	x	x	
	Phellinus rimosus (Berk.) Pilát	P	Not edible	x	x	x
	Trichaptum biforme (Fries) Ryvarden	S	Not edible	x		
	Phellinus rimosus (Berk.) Pilát	P	Not edible	x	x	
Phallaceae	*Lysurus periphragmoides* (Klotzsch)Dring	M	Edible	x		
Cerrenaceae	*Cerrena hydnoides (Sw.) Zmitr.*	S	Not edible	x	x	
Fomitopsidacea e	*Antrodia* sp.	S	Not edible			x
	Daedalea xantha (Fr.) A. Roy & A.B. De	S	Not edible	x		
Gloeophyllaceae	*Heliocybe sulcata (Berk.) Redhead & Ginns*	S	Edible		x	x
Irpicaceae	*Byssomerulius corium* (Pers.) Parmasto	S	Not edible		x	x
	Irpex lacteus (Fr.) Fr.	S	Not edible			x
Laetiporaceae	*Laetiporus sulphureus* (Bull.) Murrill	S	Edible			x

(continued)

Table 2 (continued)

Family	Species	Habit	Edibility	Sites		
				TT	BE	BPE
Panaceae	*Panus neostrigosus Drechsler-Santos & Wartchow*	S	Edible			**x**
Polyporaceae	*Cerioporus mollis (Sommerf.) Zmitr. & Kovalenko*	S	Not edible		**x**	**x**
	Daedaleopsis confragosa (Bolton) J. Schröt.	S	Not edible	**x**		
	Funalia hispida (Bagl.) M.M. Chen	S	Not edible	**x**		
	Lentinus arcularius (Batsch)Fr.	S	Not edible	**x**	**x**	**x**
	Lentinus crinitus (L.) Fr.	S	Edible		**x**	**x**
	Lentinus tigrinus (Bulliard)Fr.	S	Edible		**x**	
	Neofavolus alveolaris (DC.) Sotome & T. Hatt.	S	Edible	**x**		
	Perenniporia ohiensis (Berk.) Ryvarden	S	Not edible	**x**	**x**	
	Poria sp.	S	Not edible	**x**		
	Pseudofavolus tenuis (Fr.) G. Cunn.,	S	Not edible	**x**		
	Pycnoporus sp.	S	Medicinal		**x**	
	Pycnoporus sanguineus (L.) Murrill	S	Medicinal	**x**	**x**	
	Trametes elegans (Spreng.) Fr.	S	Not edible		**x**	**x**
	Trametes hirsuta (Wulfen) Pilát	S	Not edible	**x**	**x**	
	Trametes hirta (P. Beauv.) Zmitr., Wasser & Ezhov	S	Not edible	**x**	**x**	
	Trametes occidentalis Fr.	S	Not edible	**x**		
	Trametes variegata (Berk.) Zmitr., Wasser & Ezhov	S	Not edible	**x**		
	Trametes occidentalis Fr.	S	Not edible	**x**		
Peniophoraceae	*Peniophora albobadia (Schwein.) Boidin*	S	Not edible			**x**
	Peniophora quercina (Pers.) Cooke	S	Not edible		**x**	

(continued)

Table 2 (continued)

Family	Species	Habit	Edibility	Sites		
				TT	BE	BPE
Russulaceae	*Lactarius indigo* (Schweinitz) Kuntze	M	Edible			
	Russula brevipes Peck	M	Edible			x
	Russula cyanoxantha (Schaeff.) Fr.	M	Edible			x
	Russula delica Fr.	M	Edible		x	
	Russula mexicana Burl.	M	Edible			x
	Russula sp.	M	Not edible			x
Stereaceae	*Stereum complicatum* (Fr.) Fr.	S	Not edible		x	x
	Stereum ostrea (Blume & T. Nees) Fr.	S	Toxic	x	x	x
Diplocystidiacea e	*Astraeus hygrometricus* (Pers.) Morgan	S	Toxic		x	x
	Astraeus pteridis (Shear) Zeller	S	Not edible		x	

Edibility: C = edible, NC = Inedible, T = toxic, M = medicinal.
Habit: S = Saprotrophic, M = Mycorrhizal, P = Parasite

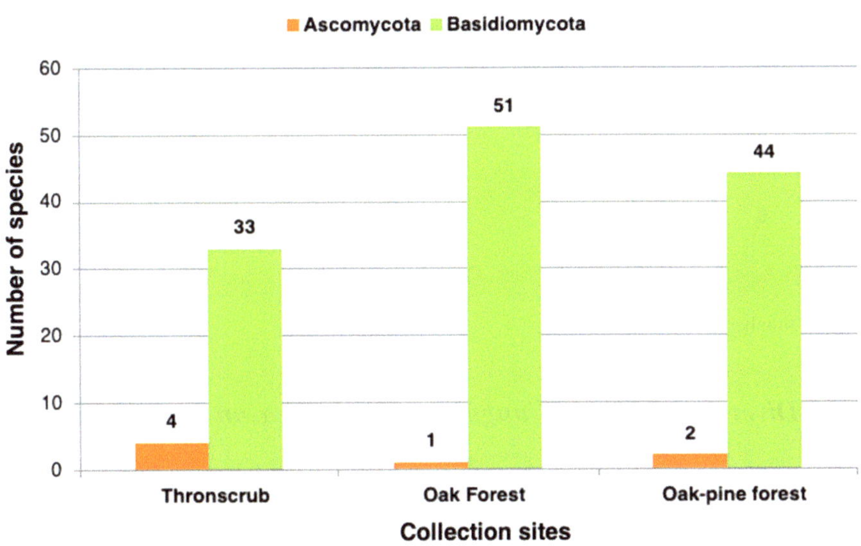

Fig. 4 Number of species of macromycetes belonging to the phylum Ascomycota and Basidiomycota recorded in each of the sites

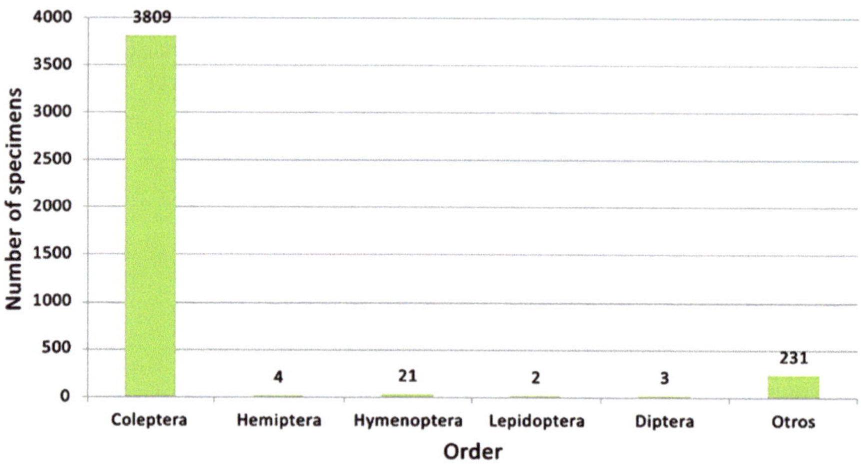

Fig. 5 Number of insect specimens collected in each order

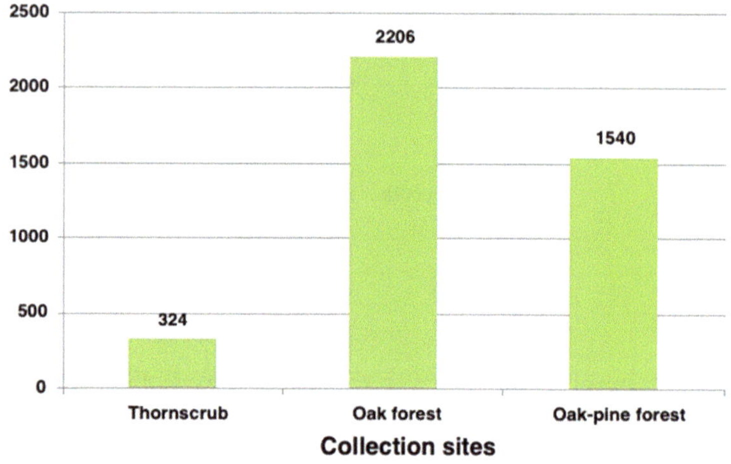

Fig. 6 Number of insects per site

3.4 *Diversity of Macro Fungal Species by Season*

The diversity of macromycetes was observed at the collection sites for each of the seasons of the year, the results showed that the diversity in autumn (September 21, 2019 - December 20, 2019) was greater with 54 species and was followed by spring (March 21, 2019 - June 20, 2019) with 36 species, winter (December 21, 2018 - March 20, 2019) with 34 species and the end of summer (June 21, 2019 - September 20, 2019) with 25 species. Only the exclusive species for each season were considered, that is, those that were recorded for each of them, and in the same

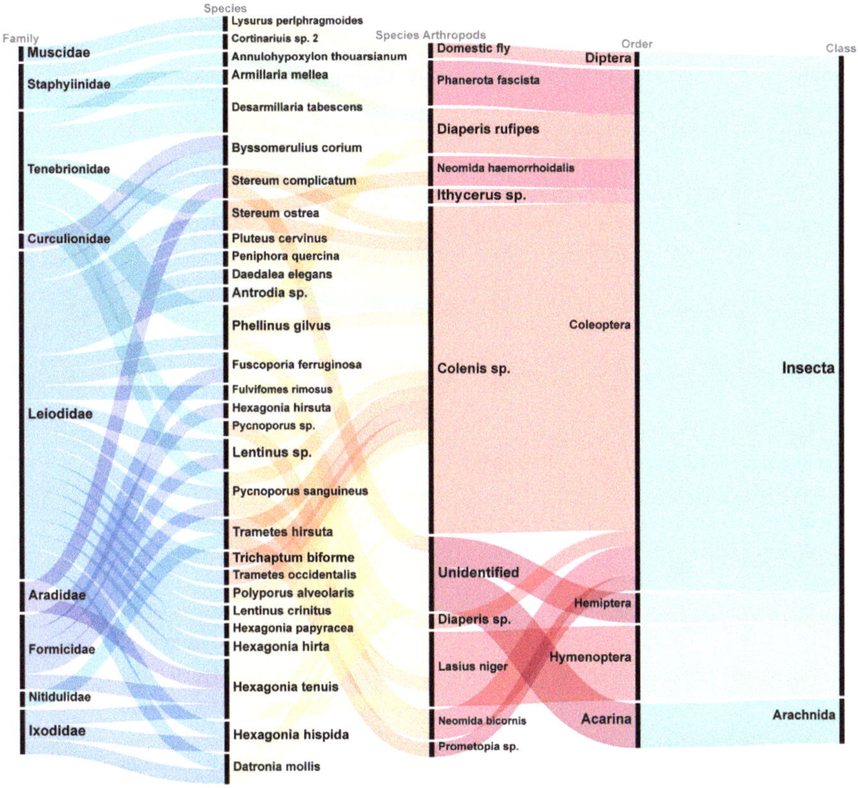

Fig. 7 Interaction between macromycetes-insect species (left) and other associated arthropods (right)

way the season presented the largest number of exclusive autumn species with 34 (i.e. *Amanita caesarea, A. pantherina, A. verna, Amanitopsis vaginata, Armillaria mellea, Astraeus pteridis, Auricularia auricula, Boletus* sp.2, *Calocybe cyanea, Cortinarius* sp. , *Cortinarius* sp.2, *Cortinarius* sp.3, *Cyathus stercoreus, Entoloma* sp., *Exidia glandulosa, Gloeophyllum sepiarium, Gymnopus erythropus, Hypomyces chrysospermus, Inocybe geophylla, Inocybe* sp., *Lactarius indigo. Limacella illinita, Lysurus periphragmoides, Neohygrocybe ovina, Omphalotus subilludens, Panellus stipticus, Pluteus* sp., *Pluteus* sp. 2, *Russula* sp., *R. brevipes, R. cyanoxantha, R. delica, R. mexicana* and *Xerocomus subtomentosus*), winter recorded 16 species (i.e. *Antrodia* sp., *Byssomerulius corium, Campanella sp., Chondrostereum purpureum, Crepidotus* sp. , *Daedalea confragosa, Dendropeniophora albobadia, Hydnochaete olivacea, Hypoxylon* sp., *Inocutis* sp. , *Lentinus strigosus, L. tigrinus, Peniophora quercina, Phellinus gilvus, Phyllotopsis nidulans* and *Trichaptum biforme*), spring 15 (i.e. *Fulvifomes* sp., *Hexagonia papyracea, Laetiporus sulphureus, Panaeolus antillarum, Polyporus alveolaris, Poria* sp., *Psathyrella condolleana, Psathyrella* sp. , *Pycnoporus* sp. ,

Table 3 Insects of each type of vegetation and food specificity in each site

Family	Species	Specificity	Associated fungi	Category	Sites 1	2	3
Araridae	ND	Polyphagy	*Pseudofavolus tenuis*	Mycetogenic	*		
		Polyphagy	*Stereum complicatum*	Mycetogenic			*
Formicidae	*Lasius niger*	Polyphagy	*Fuscoporia ferruginosa*	Mycetogenic	*		
			Trametes hirta	Mycetogenic	*		
			Pseudofavolus tenuis	Mycetogenic	*		
			Lentinus tigrinus	Mycetogenic			*
			Pycnoporus sanguineus	Mycetogenic	*		
Curculionidae	*Ithycerus sp.*	Polyphagy	*Byssomerulius corium*	Mycetogenic			*
Leiodidae	*Colenis sp*	Oligophagy	*Antrodia sp.*	Mycetobiont			*
			Byssomerulius corium	Mycetobiont			*
			Trametes elegans	Mycetobiont		*	*
			Cerioporus mollis	Mycetobiont			*
			Phellinus rimosus	Mycetobiont	*	*	*
			Fuscoporia ferruginosa	Mycetobiont	*		
			Trametes sp.	Mycetobiont	*		
			Trametes hirta	Mycetobiont	*		
			Favolus tenuiculus	Mycetobiont	*		
			Pseudofavolus tenuis	Mycetobiont	*		
			Lentinus crinitus	Mycetobiont		*	*
			Heliocybe sulcata	Mycetobiont			*
			Peniophora quercina	Mycetobiont		*	
			Phellinus gilvus	Mycetobiont		*	
			Pluteus cervinus	Mycetobiont			*
			Neofavolus alveolaris	Mycetobiont	*		
			Pycnoporus sanguineus	Mycetobiont	*	*	*
			Stereum complicatum	Mycetobiont			*
			Stereum ostrea	Mycetobiont	*		*
			Trametes hirsuta	Mycetobiont		*	
			Trametes occidentalis	Mycetobiont	*		
			Trichaptum biformis	Mycetobiont	*		

(continued)

Table 3 (continued)

					Sites		
Family	Species	Specificity	Associated fungi	Category	1	2	3
Nitulidae	*Prometopia sp.*	Polyphagy	*Trametes hirsuta*	Mycetobiont	*		
Staphylinidae	*Phanerota fascista*	Polyphagy	*Cortinarius sp.*	Mycetofilous		*	
			Armillaria mellea	Mycetofilous		*	
			Desarmillaria tabescens	Mycetofilous		*	*
Tenebrionidae	*Neomida bicornis*	Polyphagy	*Pycnoporus sanguineus*	Mycetobiont		*	*
			Pycnoporus sanguneus	Mycetobiont		*	
	Neomida hemorroidalis	Polyphagy	*Desarmillaria tabescens*	Mycetobiont		*	
			Stereum ostrea	Mycetobiont			*
	Diaperis sp.	Polyphagy	*Phellinus gilvus*	Mycetobiont		*	
	Diaperis rufipes	Polyphagy	*Desarmillaria tabescens*	Mycetobiont		*	
			Annulohypoxylon thouarsianum	Mycetobiont			*
			Phellinus gilvus	Mycetobiont		*	
Muscidae	*Musa domestica*	Polyphagy	*Lysurus periphragmoides*	Mycetofilous	*		
Ixodidae	ND	Polyphagy	*Phavolus tenuiculus*	Mycetoxenic	*		
			Cerioporus mollis	Mycetoxenic		*	
			Pseudofavolus tenuis	Mycetoxenic			*

Trametes hispida, T. occidentalis, Tremella lutescens, Truncospora ohiensis, Tulostoma sp. and *Xylaria hypoxylon*) and at the end of summer with only 7 exclusive species (i.e. *Boletus* sp., *Hygrocybe* sp., *Mycena margarita, Panaeolina foenisecii, Phaeocollybia* sp., *Pluteus longistriatus* and *Scleroderma citrinum).* Considering the sites for each of the seasons of the year, the MT site had 18 species of macromycetes in spring, 6 in summer, 13 in autumn and 16 species in winter; the site (BE) recorded 17 species in spring, 12 in summer, 20 in autumn and 16 species in winter; the site (BPE) had 8 species in spring, 11 in summer, 26 in autumn and 17 species for winter (Fig. 8). In the same way, the number of insect individuals was recorded for each of the seasons, spring showed the highest value with 1999 individuals, winter followed with 1029 individuals, autumn ha68 individuals and at the end was summer with 374 individuals of insects (Figs. 9, 10, 11, 12, and 13).

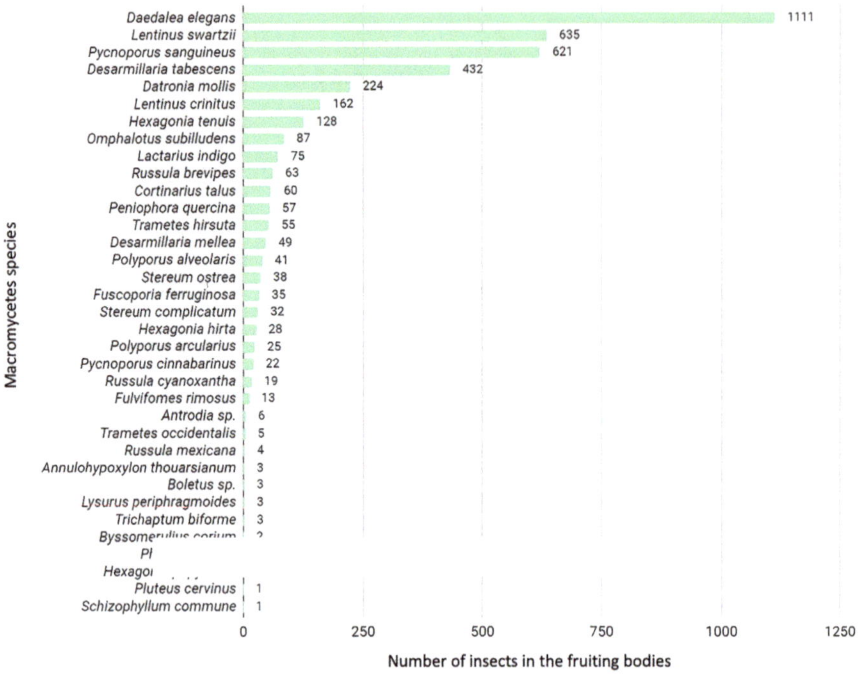

Fig. 8 Number of insect specimens in fruiting bodies of macromycetes

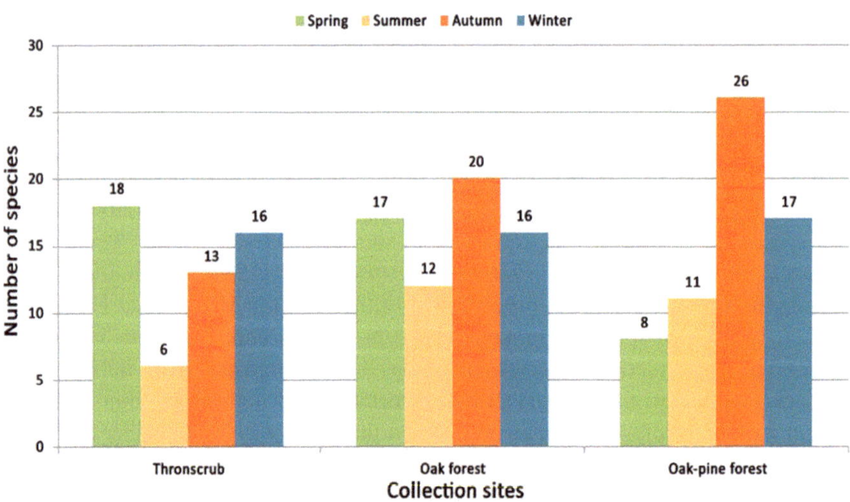

Fig. 9 Number of macromycetes species at each site per season of the year

Fig. 10 Example of diversity of macrofungal species associated with vegetation types

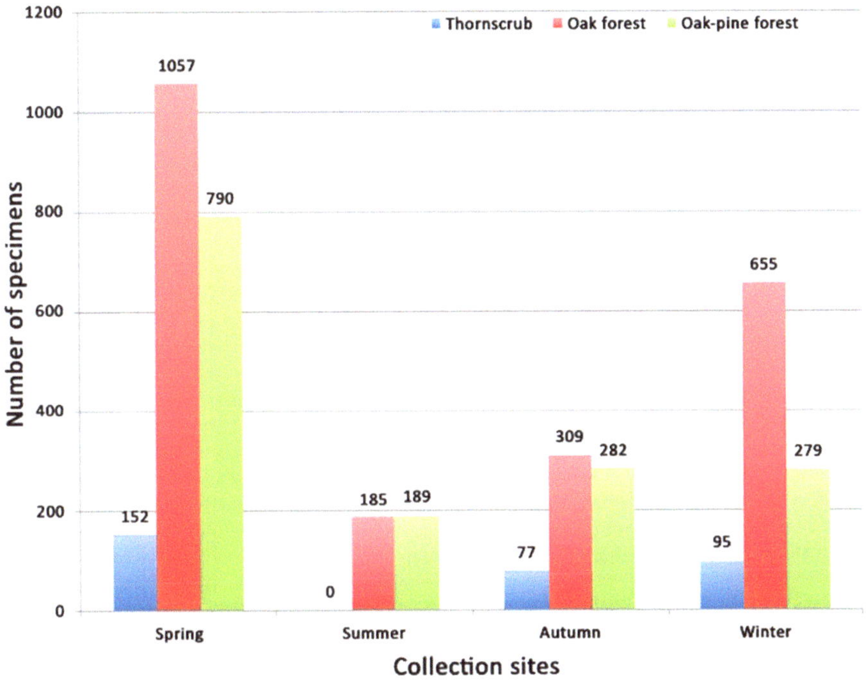

Fig. 11 Number of insect individuals present in each of the sites by season of the year

Fig. 12 Main insect species associated to sporocarps in the three localities

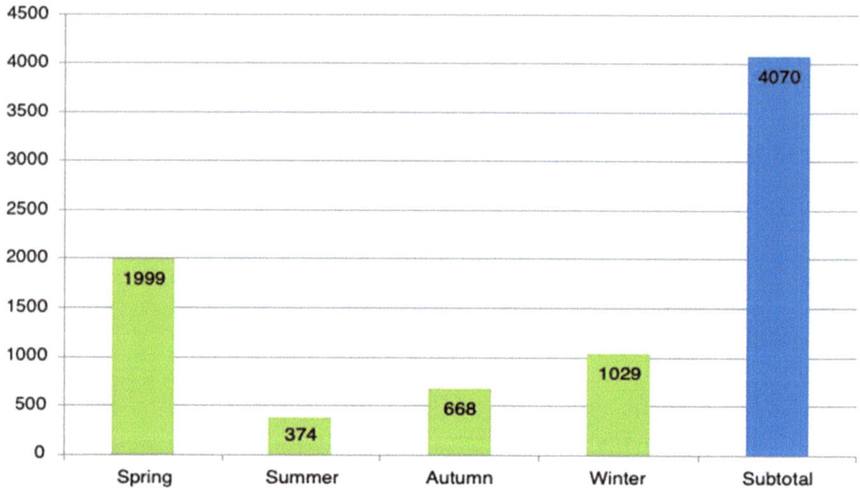

Fig. 13 Number of specimens of insects per season

3.5 Data Analysis

A similarity analysis was performed between the sites using the MVSP (Multi-Variate Statistical Package) using the Sorensen Coefficient, and the results show that site 2 and 3 (Oak Forest and Oak Pine Forest) have the m value at the top of the

Table 4 Affinity matrix with Sørensen coefficient results for the collection sites considering the macromycetes species occurring in each of them

Affinity matrix			
	Site 1 Thorn scrub	Site 2 Oak forest	Site 3 Oak-pine forest
Site 1	–	0.18	0.10
Site 2	0.82	–	0.28
Site 3	0.90	0.72	–

Table 5 Results obtained using the Shannon-Weiner Index that determines the diversity of species at each sampling site

Diversity in sampling sites		
Site 1 Thornscrubs	**Site 2** Oak forest	**Site 3** Oak-pine forest
2.50	**2.55**	**2.49**

Macromycetes species were considered

similarity value with 39.2% ($IS_S = 0.392$) and these sites are linked by site 1 with TT vegetation with a similarity value of 18.9% (ISS = 0189). An affinity matrix was made using the Sørensen Coefficient (Table 4) and the results previously obtained with the MVSP program, since the highest value occurred between sites 2 and 3 with a similarity of 28%, the lowest value was obtained for sites 1 and 3 with a similarity value of 10%.

Using the Shannon-Wiener Index, the three sampling sites were evaluated for the diversity of macromycetes species present at each site. The result of the index yields results ranging from 0.5 to 5, being values of 2–3 as ecosystems of moderate diversity, values less than 2 are considered as low diversity and values higher than 3 are ecosystems of high diversity of species.

The results showed that all three sampling sites have a high diversity, the Oak Pine Forest site had the highest value of the index with 3.59, followed by the Oak Forest site with 3.57, and at the end the MT site with 3.07. The analysis was applied in the same way to the species of macromycetes recorded by seasons of the year to know the diversity that each one had, the results obtained showed that autumn had the greatest diversity with a value of 3.84, followed by spring with 3.23, winter with 3.13 and summer with 2.90; a value of less than 3 indicates a moderate diversity for this season of the year (Table 5).

4 Discussion

According to the results of this research carried out in three types of vegetation located in the municipality of Linares, Nuevo León there is a diversity of macromycetes of 102 species, they belong to the phylum Ascomycota and Basidiomycota, the latter had the largest number of species for the three collection sites.

Considering the ecological distribution of the identified taxa, the site (BE) presented 52 species of macromycetes, being the site with the highest number of recorded species, the sites (MT) and (BPE) had 37 and 46 species respectively. The largest number of species were collected in the (BE), since it has been reported that this type of vegetation has the substrate and conditions required for the growth of most macromycetes species (Pardavé et al. 2013). This is consistent with Pardavé (1993), Pardavé et al. (2008) and Díaz et al. (2005), who mentioned that many species of fungi are widely distributed in coniferous, oak and pine forests of Mexico Insect diversity was present with 4070 individuals of Orders Coleoptera, Hemiptera, Hymenoptera, Diptera and 2 individuals of the class Arachnida, order Acarina. The Order Coleoptera had the largest number of individuals in the three vegetation types sampled with 300 individuals for the TT site, 2075 for the OF site and 1434 individuals for the BPE site. This is consistent with Barraza-Domínguez (2014), who observed that insects of the Order Coleoptera have a strong association with macromycetes species.

The association between the species of the two great Kingdoms was analyzed, observing that the family Polyporaceae recorded 13 species (i.e. *Datronia mollis, Hexagonia hirta, H. papyracea, H. tenuis, Lentinus sulcatus, L. arcularius, L. crinitus, Polyporus alveolaris, Pycnoporus cinnabarinus., P. sanguineus, Trametes hirsuta, T. occidentalis* and *Trichaptum biforme)* being the family with the largest number of species that recorded associations with insect diversity, mainly with the order Coleoptera. Considering the number of individual insects associated with each of the macromycete species, the Polyporaceae family has 1950 associated insects. On the contrary, the insect species with the highest number of associations is *Colenis* sp. belonged to the order Coleoptera, this species was present associated as a mycetobiont with 22 species of macromycetes. Amat (2007) proposed this degree of association to refer to insects whose association with the fungus is mandatory and usually depends on it to complete their life cycle. *Colenis* sp., is recorded in the sporocarps of *Antrodia* sp., *Byssomerulius corium, Daedalea elegans, Datronia mollis, Fulvifomes rimosus, Fuscoporia ferruginosa, Hexagonia papyracea, H. hirta, H. hispida, H. tenuis, Lentinus* sp., *L. crinitus, Peniophora quercina, Phellinus gilvus, Pluteus cervinus, Polyporus alveolaris, Pycnoporus sanguineus, Stereum complicatum, Stereum ostrea, Trametes hirsuta,T. occidentalis* and *Trichaptum biforme.* In the 22 species of macromycetes, galleries were observed that the Coleorptera formed to feed and deposit their waste. In the case of *Lentinus species,* 80% of their sporocarps were used as food by insects, and in some cases up to 100% of the sporoma was consumed within a period of 30 days. Despite being registered as a mycetobiont species, its specificity is analyzed as an Olygophagia, Delgado and Navarrete-Heredia (2011) recognize this level of specificity for insect species that feed on sporocarps of a few species of macromycetes, and therefore belong to genera (macromycetes) or closely related genera. The specificity pattern of *Colenis* sp. was observed and it was observed that 13 of the 22 species of macromycetes to which it is associated belong to the Family Polyporaceae, shelf fungi with soft sporocarps *(Hexagonia, Lentinus, Polyporus, Trametes)* and hard sporocarps, for example. *(Pycnoporus, Trichaptum),* this is consistent with what was previously mentioned by Delgado and Navarrete-Heredia (2011). The Sorensen Index was used to observe the similarity of the three collection sites with respect to the Macromycete species and it was observed that the OF and BPE sites had a higher similarity index (ISs=0.28), both sites share 28% of the mycobiotic diversity. On the contrary, we have the MT and BPE sites with a 10% similarity (ISs=0.10). Considering the height at which the sites are located, in the latter case the altitudinal difference is 966 m, and Núñez (1996), mentions that, in dissimilarity, height is a factor very important that influences the fruiting and formation of basidiospores. Based on the Shannon index that has been used to observe physical diversity in the different vegetation types, it was recorded as a result that the three sites have a high diversity by obtaining the values 3.07, 3.57 and 3.59 respectively.

The results reinforce the proposed hypothesis: There are differences in the composition and diversity of insect species associated with macromycetes species in different types of vegetation by season of the year, however, it will be necessary to make collections for longer periods since the phenology of some species as required, in addition to covering a larger area in the municipality. The dispersal of spores through insects is an important process that occurs throughout the year in these forests. Insect predators may also contribute to the spore dispersal process in the Halbwachs and Bässler (2015).

5 Conclusions

The results of the research show evidence of a high diversity of insect-fungus interactions across the sporocarps of macromycetes species in three vegetation types. The species of fungi involved in these interactions are required for survival and distribution in each type of vegetation in the municipality of Linares, Nuevo León. Macromycetes species are distributed in the three types of vegetation based on ecological or climatic patterns that contribute to their seasonal fruiting. The results of this research are the first report of this type of interaction for this locality in Mexico.

References

Amat-García E, Amat-García G, Henao MLG (2004) Diversidad taxonómica y ecológica de la entomofauna micófaga en un bosque altoandino de la cordillera Oriental de Colombia. Rev Acad Colomb Cienc 28(107):223–232

Amat E (2007) Caracterización de insectos micetófilos. In: Amat-García G (ed) Fundamentos y métodos para el estudio de los insectos, 1ra edn. ProOffset Editorial S.A, Bogotá, pp 135–139

Barraza-Domínguez JE (2014) Identificación de micromicetos e insectos asociados a esporocarpos en 4 diferentes tipos de vegetación en el municipio de Bocoyna, Chihuahua. Tesis de maestría. Facultad de Ciencias Forestales, U.A.N.L., Linares, Nuevo León, México

Benjamin RK, Blackwell M, Chapela IH, Humber RA, Jones KG, Klepzig KD, Weir A (2004) Hongos asociados a insectos y otros artrópodos. In: Mueller GM, Bills GF, Foster MS (eds)

Biodiversity of fungi: inventory and monitoring methods. Elsevier Academy Press, Londres, Inglaterra, pp 395–433
Delgado LL, Navarrete-HerediaJL (2011) Coleópteros micetobiontes (Insecta: Coleoptera). In Comisión Nacional para el Conocimiento y Uso de la Biodiversidad (Conabio) (Ed.). La diversidad en Veracruz: estudio de estado. Veracruz, México: Comisión Nacional para el Conocimiento y Uso de la Biodiversidad, Gobierno del Estado de Veracruz, Universidad Veracruzana, Instituto de Ecología, A. C.
Díaz M, Marmolejo J, Valenzuela R (2005) Flora Micológica de Bosques de Pino y Pino-Encino en Durango. Ciencia UANL 8(3):362–369
García-Hernández J, Jurado E (2008) Caracterización del matorral con condiciones prístinas en Linares N.L., México. Ra Ximhai 4(1):1–21
Halbwachs H, Bässler C (2015) Lo que el viento se llevó una revisión sobre las basidiosporas de lamellate agarics. Micosfera 6(1):78–112. https://doi.org/10.5943/mycosphere/6/1/10
Hanski, I. (1989). Fungivory: Hongos, insectos y ecología. En Wilding, N., Collins, N.M., Hammond, P.M., & Weber, J. F. (Eds.). Interacciones insecto-hongo. Londres, Inglaterra: Academic Press.
Kirk PM, Cannon PF, Minter D, Stalpers. (2008) Diccionario de los hongos, 10ª edn. CABI Publishing, Inglaterra
Lawrence, J. F. (1989). Micofagia en los coleópteros: Estrategias de alimentación y adaptaciones morfológicas. En N. Wilding, N.M Collins, P. Hammond & J. F. Webber (Eds.), Interacciones insectos-hongos. (pp. 1-23). New York: Academy Press.
Lodge DJ, Ammirati JF, O'Dell TE, Mueller GM, Huhndorf SM, Wang CJ et al (2004) Macrohongos terrestres y lignícolos. En: Mueller GM, Bills GF, Foster MS (eds) Biodiversidad de hongos: métodos de inventario y monitoreo. Academic, New York, pp 127-158
Núñez M (1996) Fructificación de Polyporaceae s.I. (Basidiomycotina) a lo largo de un gradiente de altitud y humedad en el Área de Conservación Guanacaste. Rev Ecol Trop 12:893–898
Pardavé L (1993) Macromicetos de Sierra Fría. Investigación y Ciencia 10:24–29
Pardavé L, Flores L, Franco V, Robledo M (2008) Hongos y Líquenes del Estado de Aguascalientes. In: La Biodiversidad de Aguascalientes México. CONABIO, IMAE, UAA
Pardavé L, Flores L, Castañeda R, Ruiz., V. (2013) Diversidad de Macromicetos en el Municipio de San José de Gracia. Aguascalientes Revista Investigación y Ciencia 57:11–18
Putman RJ (1994) Ecología comunitaria, 1ra edn. Kluwer Academic Publishers, Londres, Inglaterra

Interactions Between Macrofungals and Insects in the Oak and Pine Forest in the Municipalities of Iturbide and Galeana, Nuevo León

Edwin Garza López, Fortunato Garza-Ocañas, Humberto Quiroz Martínez, Gerardo Cuellar Rodríguez, and José Isidro Uvalle Sauceda

Abstract The present study is focused on the association of insects to sporocarps of macromycetes collected in two municipalities of Nuevo Leon. Kurskall – Wallis statistic test was used to establish the association level of the organisms. Vegetation characterization was made on the studied sites in a way to make a precedent over the abiotic variables that affect the presence of the macromycetes. The dominant species were *Juniperus flaccida*, *Quercus* spp., *Pinus pseudostrobus* for both sites. 75 fungus species were identified distributed on 36 families and 58 genera. The family Tricholomateace was the most abundant with 14 species distributed in 9 genera. Regarding insects 44 were identified from sporocarps of macromycetes, they belonged to the orders Coleoptera, Himenoptera, Lepidoptera and Collembola. Abiotic variables were studied to study their relationship on the presence of macromycetes, an analysis of climatic variables was made of the Altavista meteorologic station of INIFAP determining that the precipitation that occurred in 2016 was very favorable for the fructification of the fungal species.

Keywords Fruiting bodies · Insects · Temperate forests · Interactions

1 Introduction

Every way of life, no matter how small, is unique. On respect for all of them depends on the balance of the Earth (Hongos de La Tierra Declaration of Córdoba 2007). Mexico's terrestrial biological diversity comprises a wide variety of landscapes and plant communities that cover the country's territory (Challenger and Soberón 2008).

E. G. López · F. Garza-Ocañas (✉) · G. C. Rodríguez · J. I. U. Sauceda
Facultad de Ciencias Forestales, Universidad Autónoma de Nuevo León, Linares, Nuevo León, Mexico

H. Q. Martínez
Facultad de Ciencias Biológicas, Universidad Autónoma de Nuevo León, Laboratorio de Entomología, San Nicolás de los Garza, Nuevo León, Mexico

F. Garza-Ocañas (ed.), *Sustainable Management of Natural Resources*, Earth and Environmental Sciences Library, https://doi.org/10.1007/978-3-031-33394-1_14

In Nuevo León there are different types of vegetation where pine, oak and pine-oak forests stand out (Vargas 1999). The natural temperate forests that cover an important part of the semi-dry and temperate zones of northern Mexico have been affected by anthropogenic actions of various kinds. Some forest activities such as thinning, natural and prescribed fire, grazing and recreation can change the character of the forest soil and the mushroom community (Quiñónez-Martínez et al. 2013). Fungi as decomposers par excellence play the role of soil generation, which allows the establishment of plant species who in turn will attract fauna and fauna through mycophagy or the simple rubbing of sporocarps will function as dispersers of fungi to re-enter this cycle. Without fungi ecosystems could not be established (Harley 1989; Stamets 2005). Some fungi form ectomycorrhizas in the temperate forests of the Sierra Madre Oriental and these are associated with species of forest importance in Nuevo Leon. Some studies have been carried out in which the ectomycorrhizal fungi associated with different plant communities are described, such as *Quercus*, *Pinus*, *Abies*, *Picea*, *Pseudotsuga*, *Cupressus*, *Juniperus* and *Taxus* (Garza 1986). The spread of fungi is so important, that if one adds up all the sporocarps produced in a single acre of soil. Of forest, the total number of spores is literally astronomical, as incomprehensible as the number of stars in the universe (Maser et al. 2008). Fungi have different dispersal strategies with which they spread in new areas within ecosystems (Dighton 2003). There are different methods of dispersal of spores, one of them is the concentration of spores in the excreta of animals that feed on truffles being an advantageous method to form new mycorrhizae with trees (Maser et al. 2008). Mycophagia has played a fundamental role in the evolution of the reproductive systems and dispersal mechanisms of hypogeal and epigeal fungi (Castillo-Guevara et al. 2012). Preferential mycophages include those organisms that normally consume fungi to a greater degree than all other types of food available. Others such as casual or opportunistic mycophages include species that consume truffles when looking for other food objects or alternatively consume truffles when preferred food sources are temporarily unavailable (Maser et al. 2008). In the specific case of insects, we can appreciate symbiosis, so that the appearance of fungi in forests not only appears as a refuge or opportunistic food source but also leads to an even more intimate relationship; it could be said that organisms work for mutual survival in association, mutualism, involves the benefit of both partners in an association. The primary benefit of the fungus is probably the safe dispersal (apart from wind dispersal) of the spores, and their inoculation into new habitats suitable for their growth.

As an example, the tunnels of some species of beetles help a rapid penetration of the mycelium, once established, the fungi benefit from the activities of the beetles. These organisms are helped in ways that are poorly understood to maintain dominance of the ambrosia fungus in galleries, as well as to prevent or prevent the growth of other fungi (Beaver 1989). The mycophagy strategies presented by fungi in conjunction with animals (Maser et al. 1988), not only does the intake of the sporocarp allows the mobility of the fungus, but abiotic factors also have an important role in dispersal; as an example, the spores of the Mesophellia are extremely hydrophobic, so the impact of the raindrops that hit the mass of spores will cause them to fly into the air. Other spores cling to the legs and hair of the consumer e.g., field rat and truffle and the spores are dispersed by the animal that moves in the forest and releases its secretions full of spores ready to germinate (Maser et al. 2008). Relatively few mammal species around the world derive part of their nutritional requirements from fungal sporocarps and are therefore considered obligate mycophages. Forced mycophagia represents either the coevolution of the consumed fungus or an adaptation as strong as virtual coevolution. In the case of mammals, such mycophages require habitats in which fruiting bodies are available throughout the year or if the climate oscillates the production of fruiting bodies for a short period, an alternative food source is available as a supplement (Maser et al. 2008). The abundance of diversity and composition of fungi depends largely on the arrangement of resources available to fungi that inhabit thick timber remains, while fungal species that develop in fine timber remains to depend heavily on the microclimate (Bässler et al. 2010). Most macromycetes produce fruiting bodies during the fall, but some have adapted to fruiting in spring, it has been found that there is a correlation in the climatic conditions of one year and the fruiting of the next, indicating that the mycelium that is under the ground is influenced by climatic conditions for a longer period before fruiting. Global warming has led to early spring fruiting in European countries (Köppen, 1918; Kauserud et al. 2010). Thus, we can visualize in a not very deep way how these interactions are so broad, complex, and specific for certain species, so it is necessary to study them to try to understand how their relationships are and not to divert the value of importance that the relationship between insects with the sporocarps of macrofungi has in the forest. The lack of knowledge of the associations of fungal-insect interaction gives a guideline to propose this research.

2 Methodology

2.1 Location of the Study Area

The study was conducted in the south of the state of Nuevo León in the municipalities of Iturbide and Galeana where the altitude ranges between 700 and 3000 meters above sea level (INEGI 2009). Within this municipality is the Iturbide Ecological Campus of the Autonomous University of Nuevo León, better known as "Bosque Escuela". In addition to the plots within the Campus, data were taken from 3 plots established in the Ejido Puerto de Reyna 11.5 kilometers away adjacent to the municipality of Iturbide, where there is a deployment of hills with conditions very similar to those of the school Forest to establish a comparison between the sites.

2.2 Soil, Climate, and Vegetation

Within the municipality of Iturbide, the dominant soils are: Leptosol (95.5%), Calcisol (2.8%) and Kastañozem (1.7%). As far as geology is concerned, there are sedimentary rocks: Limestone (49%), limestone-shale (36%), shale (10%), conglomerate (2.8%), gypsum (1%) and sandstone-conglomerate (0.2%) (INEGI 2009). With respect to the climate of the study area has a temperature range of 12–22 °C its precipitation range is from 500 to 800 mm and the climate is divided into: Semi-dry semi-warm (54%), Temperate subhumid with rains in summer, lower humidity (34%), Semi-warm subhumid with rains in summer, lower humidity (11%) and Temperate subhumid with rains in summer of average humidity (1%). The climatic units of the two sites are represented by the (INEGI 2005) as Temperate subhumid for the plots established in Puerto de Reyna and Semi-dry semi-warm for the school Forest. The types of vegetation found in Iturbide are forest (67%), scrub (28%) and grassland (1%). The percentage belonging to forest is composed of oak forests, mixed oak and pine forests and pine forest, only 4% of the territory of the municipality is destined to the use of agricultural land, while in the municipality of Galeana most of the territory is scrub (55%), forest (23%), grassland (6%) and agriculture (15%) (INEGI 2009) (Fig. 1).

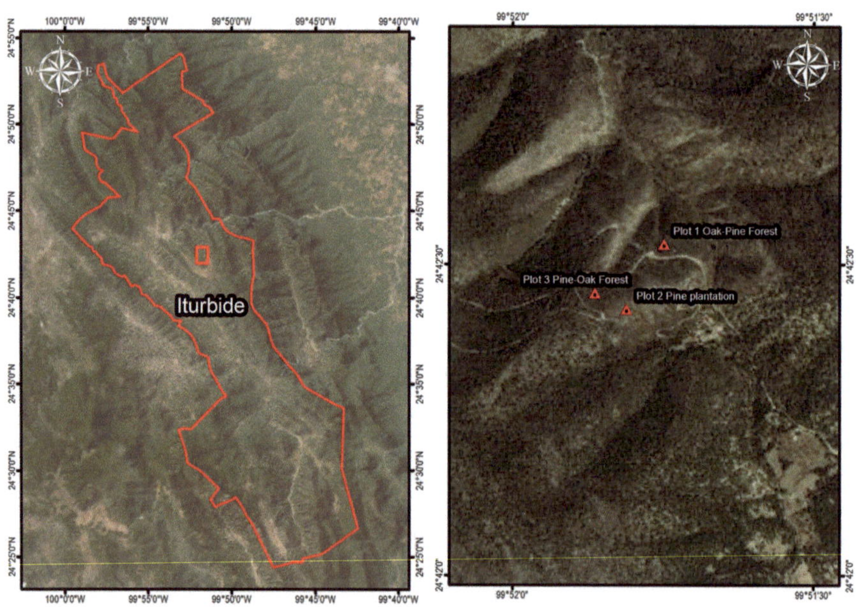

Fig. 1 Location of the study area

2.3 Characterization of Vegetation

Sampling was carried out at 2 sites with 3 types of vegetation: oak-pine forest, pine forest, and pine-oak forest. In these sites, 3 plots of 10 m × 15 m (150 m^2) were established to describe the vegetation, the coverage, dominance, and frequency with which the plant species appeared were recorded (Repetto Giavelli 2012) (Fig. 2).

2.4 Collection of Macromycetes and Insects

Collections were carried out for two years during the rainy season. In each transect the sporophores of the species were collected, this was preserved in bags of waxed paper with a label with the data of field, transect and area. The specimens were taken to the laboratory, a part of them was dehydrated and labeled for later identification other specimens were placed in the incubation chambers to later collect the insects that emerged, these were collected and processed for photography and placed in jars with 70% alcohol Márquez Luna (2005).

2.5 Identification of Macromycetes and Insects

For the identification of macromycetes, the studies by (Singer and Guzman 1978, Ostry et al. 2010; O'Brien, J.G., **2010**; Fergus and Fergus 2003) were used as well as queries from websites such as www.mycobank.org (Mycobank database n.d.) and www.indexfungorum.org (Index Fungorum n.d.).

Fig. 2 Establishment of collection plots

2.6 Insect Conservation

The identification of the insects was carried out in the Entomology Laboratory of the Faculty of Biological Sciences of the UANL the specimens were prepared for photography in a camera mounted to the stereoscope. The insects were collected from the sporocarps and then placed in containers with 70% alcohol to preserve them as mentioned by Márquez Luna (2005).

2.7 Statistical Analysis

To perform the statistical analysis between the macromycetes and insects that were presented, simple correspondence analysis was performed, looking for the statistical dependence of insects with mushrooms. The data collected in databases were agglomerated to be able to process them in different statistical packages, determining the appropriate indicators for the analysis of biodiversity, as well as the selected abiotic variables that may have a direct relationship with the appearance of the mushrooms of the selected sites. The data was processed under the statistical software SPSS (IBM Corp. 2016).

Chi-Square Test

Chi-square-proof was chosen to determine the statistical association of macromycetes with the presence of insects. The Chi-square test which is developed with qualitative variables, (presence, absence). Likewise, the results observed in an investigation with a set of theoretical results, under the assumption that the variables were independent. The difference between the observed and expected results is summarized in the value adopted by the X2 statistic, which has an associated p-value, below which the hypothesis of independence of the variables is accepted or rejected (Cerda and Villarroel Del 2007).

The test formula of X^2

Where:

X^2 = Chi-squared

$\sum$ = Sum of events

O = Observed events

E = Estimated events

$$X^2 = \sum_{i=1}^{n} \left(\frac{(O_i - E_i)^2}{E_i} \right)$$

Kruskal Wallis Test

The Kruskal Wallis test was chosen to group the different types of vegetation (Oak/Pine Forest, Pine Forest/Oak, Pine Plantation), as it is an extension of the Mann-Whitney test for more than two groups. The Kruskal-Wallis test is suitable when the data have a natural order, that is, when to make sense of them they have to be ordered or when the conditions for applying an analysis of variance are not satisfied. This test evaluates whether the sums of ranges differ from each other so much, that it turns out not it is reasonable to consider that they come from samples that were randomly selected from the same population (Rodrigo 2016).

$$K=(N-1)\frac{\sum_{i=1}^{g}n_i(\overline{r_i}-\overline{r})^2}{\sum_{i=1}^{g}\sum_{j=1}^{n_i}\left(r_{ij}-\overline{r}\right)^2}$$

Where:

n_i is the number of observations in the group i

r_{ij} is the range (between all observations) of the j observation in the group i

N is the total number of observations between all groups

Sørensen Index

The formula original of Sørensen was intended to apply data of presence/absence, and it is defined as follows:

$$QS=\frac{2C}{A+B}=\frac{2|A\cap B|}{|A|+|B|}$$

- QS is the similarity quotient and varies from 0 to 1
- A and B are the number of species in samples A and B
- C is the number of species shared by the two samples

This expression easily extends to abundance rather than the presence/absence of species (Sørensen 1948).

3 Results

3.1 Characterization of Vegetation

Regarding the characterization of the school forest, a very high presence of *Juniperus flaccida* was found, in the 3 established plots with an important value index of 19.9% followed by *Pinus pseudostrobus* (15.5%) and Quercus spp, (16.1%). The species that presented the highest frequency were *Juniperus flaccida, Pinus pseudostrobus, Quercus canbyi, Quercus polymorpha, Astragalus hypoleucus, Croton incanus, Arbutus xalapensis*, indicating the condition of the forest. The characterizations of vegetation in the two sites come to highlight interesting data, in the school forest the species with the highest importance valuc index (IVI) was *Juniperus flaccida* with 19.9% of the total species. This can be caused by the great coverage that the species presented on the sites. However, the presence of *Pinus pseudostrobus* and *Quercus* spp somewhat complete the initially proposed picture that the plots were established in Pine Forests – Oak and Oak Forests – Pine. Many of the established fungal species obey a pattern of presence of these species in vegetation (Garza 1986). The site established in the Ecological Campus Forest School of the UANL, has been to some extent restricted from the affectation of anthropocentric activities for a little more than 35 years, the establishment of this area protected by the University allowed the property to be exempt from domestic livestock throughout its facilities. When an ecosystem loses functionality and degrades, physical and floristic changes occur in the vegetation, observing a decrease in total coverage (Gaitán et al. 2009). In Site 2 established in the Ejido Puerto de Reyna, Galeana, N.L., the forest presents characteristics very similar to those of the school Forest in terms of structure, the sites are at a distance of 11.5 kilometers relatively close. However, the specific conditions of the mountain range where the forest of Puerto de Reyna is located, probably establish it as a refuge area for the plant and fungal species. The refuge zones are established by the climatic cycles and glaciations of the ice age, these temperature clades allowed in their time the establishment of bottlenecks that are associated with expansions of plant populations from an equatorial refuge to temperate habitats and contractions back to these refuge areas when the temperature dropped (Gadd et al. 2007). The site is isolated by the formations of the Sierra Madre Oriental and the sites are in a valley within the Ejido. Considering these aspects of Site 2 explains the findings of the macromycetes unique to this site. As for the characterization of this site, the species with the highest index of importance value was *Bauhinia forficata*, highlighting that this species is the most abundant (50.7%), since it presented a large amount of regeneration due to this very dominance of the species was very low (2.9%), this species is consumed by ruminants as mentioned by Pereira and José (2015) through histological evaluation and visual observation of livestock.

Unlike the School Forest, the site of Puerto de Reyna has frequent grazing of domestic livestock, mainly cattle, which obviously affects the composition of the vegetation and explains to some extent the great regeneration of this species in the plots of this site. Another favorable comparison that we can make in the Site of Puerto de Reyna is the presence of *Quercus* spp. and *Pinus pseudostrobus* with the highest relative dominances (26.4% and 22.3%), which like Site 1, establishes it with the ecosystems of Oak Forest – Pine and Pine Forest – Oak. In the species diversity of Site 1 27 plant species were found while in Site 2 only 23 species were found. Similarly, the greater diversity of species shows a relationship with the results of the macromycetes found at the sites.

3.2 *Species of Macromycetes*

In total, 75 species were identified, which are distributed within 36 families and 58 genera. The family Tricholomateace was the most abundant with 14 species spread over 9 genera, followed by Agaricaceae with 5 species within 5 genera. Table 1 shows the habit of the macromycetes collected during the investigation as

Table 1 Species in the sites of collection, habit and edibility

Family	Species	Site 1	Site 2	Habit	Edibility
Agaricaceae	*Agaricus sp.*	X	X	Saprobe	–
	Lepiota sp	X	X	Saprobe	–
	Leucoagaricus rubrotinctus (Peck) Singer		X	Saprobe	Unknown
	Leucocoprinus fragilissimus (Ravenel ex Berkeley & M.A. Curtis)		X	Saprobe	Unknown
	Lycoperdon echinatum Pers.	X		Saprobe	Edible (young fruit) (1)
Amanitaceae	*Amanita fulva Fr.*	X		Mycorrhizal	Edible/easily mistaken
	Amanita vaginata (Bull.) Lam.	X		Mycorrhizal	Edible without reference
	Amanita verna (Bull.) Lam.	X		Mycorrhizal	Mortal Toxic
	Amanita virosa Bertill.		X	Mycorrhizal	Mortal Toxic
Amylostereaceae	*Artomyces pyxidatus (Pers.) Jülich*	X		Saprobe	Edible without reference
Auriscalpiaceae	*Auriscalpium vulgare Gray*	X		–	–
Boletaceae	*Boletus rubellus Krombh.*		X	Mycorrhizal (2)	Edible non culinary
	Cyanoboletus sp.		X	–	–
	Frostiella russellii (Frost) Murrill, Mimeo	X		Mycorrhizal	Edible
	Strobilomyces floccopus (Vahl) P. Karst.	X		Mycorrhizal	Edible (3)
Cantharellaceae	*Cantharellus cibarius Fr.*	X	X	Mycorrhizal	Edible
	Cantharellus igniformis.	X	X	Mycorrhizal	Edible
Catathelasmataceae	*Callistosporium sp. Singer*	X		Saprobe	–
Coprinaceae	*Coprinus comatus (O.F. Müll.) Pers.*	X		Saprobe	Edible
	Coprinus lagopus (Fr.) Fr.	X		Saprobe	Non Edible
	Coprinus plicatilis (Curtis) Fr.		X	Saprobe	Edible non culinary
Cortinariaceae	*Cortinarius phoeniceus (Bull.) R. Maire*	X	X	Mycorrhizal (4)	Toxic
Cyphellaceae	*Chondrostereum purpureum (Pers.)*	X		Parasite -Saprobe	Non Edible
Dacrymycetaceae	*Dacryopinax spathularia (Schwein.) G.W. Martin*	X		Saprobe	Edible

(continued)

Table 1 (continued)

Family	Species	Site 1	Site 2	Habit	Edibility
Entolomataceae	*Entoloma lividum (Bull.) Quél*	X	X	Mycorrhizal	Toxic / Could be mortal
Geastraceae	*Geastrum saccatum Fr.*	X		Saprobe	Without culinary interest
Gloeophyllaceae	*Gloeophyllum sepiarium (Wulfen) P. Karst.*	X		Saprobe	Medicinal
Gomphidiaceae	*Chroogomphus vinicolor (Peck) O.K. Mill.*	X		Mycorrhizal	Edible
Helvellaceae	*Helvella crispa Bull.*	X		Mycorrhizal	Edible not recommended
Hydnangiaceae	*Laccaria laccata (Scop.) Cooke*	X	X	Mycorrhizal	Edible not recommended
Hygrophoraceae	*Hygrophorus coccinea*	X		Mycorrhizal	Edible
Hymenochaetaceae	*Inonotus P. Karst.*	X		–	–
Inocybaceae	*Crepidotus mollis*	X	X	Saprobe	Non Edible
	Inocybe calamistrata (Fr.) Gillet	X	X	Mycorrhizal	Toxic (5)
Marasmiaceae	*Macrocystidia Joss.*		X	–	–
	Marasmius androsaceus (L.) Fr.	X	X	Saprobe	Without culinary interest
	Marasmius maximus Hongo	X	X	Saprobe	Edible
	Marasmius oreades (Bolton) Fr.	X		Saprobe	Edible
Meruliaceae	*Merulius incarnatus Schwein.*	X		Saprobe	Non Edible
Mycenaceae	*Mycena epipterygia (Scop.) Gray*	X	X	Saprobe	Non Edible
	Xeromphalina cauticinalis (Fr.) Kühner & Maire	X	X	Saprobe	Without culinary interest
Omphalotaceae	*Gymnopus dryophilus (Bull.) Murrill*	X	X	Saprobe	Edible
	Gymnopus fusipes (Bull.) Gray	X	X	Parasite -Saprobe	Edible
Polyporaceae	*Coriolus versicolor (L.) Quél.*	X		Saprobe	Medicinal
Paxillus	*Paxillus panuoides (Fr.) Fr.*		X	Saprobe	Non Edible
Pleurotaceae	*Hohenbuehelia petaloides (Bull.) Schulzer*		X	Saprobe	Edible
Psathyrellaceae	*Coprinopsis sp.*	X		–	–
	Psathyrella sp.		X	–	–

(continued)

Table 1 (continued)

Family	Species	Site 1	Site 2	Habit	Edibility
Russulaceae	*Lactarius indigo (Schwein.) Fr.*	X		Mycorrhizal	Edible
	Lactarius uvidus (Fr.) Fr.	X	X	Mycorrhizal	Without culinary interest
	Russula emetica (Schaeff.) Pers.	X	X	Mycorrhizal	Toxic
Sarcoscyphaceae	*Sarcoscypha occidentalis. (Schwein.) Sacc.*	X		Saprobe	Non Edible
Stereaceae	*Stereum complicatum (Fr.) Fr.*	X		Saprobe	Non Edible
	Stereum ostrea (Blume & T. Nees) Fr.	X		Saprobe	Non Edible
Schizophyllaceae	*Schizophyllum commune Fr.*	X		Parasite – Saprobe	Edible
	Schizophyllum umbrinum Berk.	X		Parasito – Saprobe	–
Strophariaceae	*Deconica coprophila (Bull.) P. Karst.*	X		Saprobe	Toxic
	Gymnopilus aeruginosus (Peck) Singer	X		Saprobe	Non Edible/ Psychoactive
	Stropharia semiglobata (Batsch) Quél.	X		Saprobe	Without culinary interest/ potencial toxic
Sclerodermataceae	*Scleroderma areolatum Ehrenb.*	X		Mycorrhizal	Non Edible
Suillaceae	*Suillus granulatus (L.) Roussel*	X	X	Mycorrhizal	Edible by preparation, possible laxative
Tricholomataceae	*Clitocybe alba (Bataille) Singer,*	X	X	Saprobe	Non Edible
	Clitocybe gibba (Pers.) P. Kumm.	X		Saprobe	Edible
	Collybia butyracea (Bull.) P. Kumm.	X		Saprobe	Edible
	Hygrocybe coccinea (Scchaeff.) P. Kumm.	X		Mycorrhizal	Edible
	Lepista sórdida (Schumach.) Singer		X	–	Edible
	Lepista nuda (Bull.) Cooke	X	X	Saprobe	Edible
	Leucopaxillus albissimus (Peck) Singer	X		Saprobe	Non Edible
	Leucopaxillus gentianeus (Quél.) Kotl.	X		Mycorrhizal	Non Edible

(continued)

Table 1 (continued)

Family	Species	Site 1	Site 2	Habit	Edibility
	Melanoleuca sp.		X	–	–
	Pseudoclitocybe sp.		X	–	–
	Resupinatus applicatus (Batsch) Gray	X		Saprobe	–
	Tricholoma ustale P. Kumm.	X		Mycorrhizal	Toxic
	Tricholoma sulphureum (Bull.) P. Kumm.	X		Mycorrhizal	Toxic
	Tricholoma terreum (Schaeff.) P. Kumm.	X		Mycorrhizal	Edible by preparation

1 (Hanlin 2004) 2 (Region 1980) 3 (Sturgeon 2018) 4 (Marks 2012) 5 (Flores Cavada et al. 2018)

well as their edibility. Of the 74 species that identified the macromycetes of: *Agaricus* sp., *Lepiota* sp., *Cantharellus cibarius*, *Cantharellus igniformis*, *Cortinarius phoeniceus*, *Entoloma lividum*, *Laccaria laccata*, *Crepidotus mollis*, *Inocybe calamistrata*, *Marasmius androsaceus*, *Marasmius maximus*, *Mycena epipterygia*, *Xeromphalina cauticinalis*, *Gymnopus dryophilus*, *Gymnopus fusipes*, *Lactarius uvidus*, *Russula emetica*, *Suillus granulatus*, *Clitocybe alba* and *Lepista nuda* (20) were collected at both sites. The exclusive species of Site 1 (41) "School Forest" were: *Lycoperdon echinatum*, *Amanita fulva*, *Amanita vaginata*, *Artomyces pyxidatus*, *Auriscalpium vulgare*, *Frostiella russellii*, *Strobilomyces floccopus*, *Callistosporium* sp., *Coprinus comatus*, *Coprinus lagopus*, *Chondrostereum purpureum*, *Dacryopinax spathularia*, *Geastrum saccatum*, *Gloeophyllum sepiarium*, *Chroogomphus vinicolor*, *Helvella crispa*, *Hygrophorus coccinea*, *Inonotus* sp., *Marasmius oreades*, *Merulius incarnatus*, *Coriolus versicolor*, *Coprinopsis* sp., *Lactarius indigo*, *Sarcoscypha occidentalis*, *Stereum complicatum*, *Stereum ostrea*, *Schizophyllum commune*, *Schizophyllum umbrinum*, *Deconica coprophila*, *Gymnopilus aeruginosus*, *Stropharia semiglobata*, *Scleroderma areolatum*, *Clitocybe gibba*, *Collybia butyracea*, *Hygrocybe coccinea*, *Leucopaxillus albissimus*, *Leucopaxillus gentianeus*, *Resupinatus applicatus*, *Tricholoma ustale*, *Tricholoma sulphureum*, *Tricholoma terreum*. The exclusive species of Site 2 (i.e.13) "E.g. Port of Reyna" were: *Leucoagaricus rubrotinctus*, *Leucocoprinus fragilissimus*, *Amanita verna*, *Amanita virosa*, *Boletus rubellus*, *Boletellus* sp., *Coprinus plicatilis*, *Paxillus panuoides*, *Hohenbuehelia petaloides*, *Psathyrella* sp., *Lepista sordida*, *Melanoleuca* sp., *Pseudoclitocybe* sp. The greater diversity found in site 1 can be due to two reasons, the first was the greater number of collections that were made in Site 1 in 2016 and 2017 as the main area of study of the research, in contrast to the establishment of the plots in Site 2, there was only collection of macromycetes in 2017, this with the intention of establishing a contrast with the findings of the first site. The second reason is that the vegetation at site 1 is to some extent protected from grazing damage, which could favorably affect the appearance of macromycetes as there is no trampling of domestic livestock. Despite the differences the sites share 20 species some of these obeying the presence of vegetation, this is because the roots of most species in plant communities are colonized by symbiont fungi that tend to form mycorrhizae, which play a critical role in the assimilation of soil nutrients and consequently to the nutrition of the plant (Smith et al. 2008). Among the macromycetes found in Site 2, *Boletellus* sp. and *Leucoagaricus rubrotinctus*. The genus *Cyanoboletus* has several species in Mexico but the one presented here does not correspond to any of those known in the literature, so it might be a new species. For these purposes, DNA samples were taken and analyzed for the correct determination of the species. Specimens from Site 2 Ejido Puerto de Reyna were growing next to *Quercus canby*.

3.3 Relationship Fungus – Insects

A total of 107 specimens of macromycetes were collected, of which 56 had the presence of insects, to establish a relationship between the fungi found in the different types of ecosystems, the Chi-square test was carried out, which is developed with qualitative variables, these variables (presence, absence) yielded a test value of 107, in which 100% of the boxes expected a recount less than five. With this statistical test we can statistically define an association between sporocarps and the presence of insects with a p value < 0.05.

3.4 Insect-Vegetation Relationship

To define if there was a difference in any of the ecosystems where the sporocarps were collected with the insects, the Kruskal Wallis Test was carried out, where a value of $p \geq 0.05$ was obtained, so it can be statistically deferred that there is no significant difference between the appearance of insects in macromycetes with the types of ecosystems where they were collected.

3.5 Identification of Species of Macromycetes and Associated Insects

The reality is that many of the macromycetes that are sought in the field have very high probabilities of containing some other organism within them, it is necessary to introduce a little to the world of mycology to understand that fungi as a food resource and refuge are very appreciated. One of the likely questions of what happens to the interdependence of some of the insect species when conditions are not favorable for the fruiting of fungi is what happens to insects? Does the fruiting of the fungus disappear and does the insect disappear? The response to how environmentalists see it is always to evade the unfavorable conditions of "the here and now" – dispersion and diapause (Hanski 1989). Insects that present diapause are not uncommon to use resources that are unpredictable and temporary. An example of this is the insects that develop in the seeds and cones of some conifers that often present dramatic changes in availability year after year (Annila 1982). The insects found in this research probably do not use this strategy because in the forest different specimens of insects were found in different fungi, not being strictly related to a host fungus these can participate in different interaction dynamics and take advantage of more resources available in the ecosystem. To determine the degree of association between sporocarps and insects it is important to consider the roles played by each of the individuals in the ecosystem, Bruns (1984), distinguishes between 4 higher trophic levels of insects that develop in fungi; Primary fungivores, secondary fungivores, detritivores and predators. It is likely that Bruns's (1984) classification is related to the evolution of mycophagia in insects, detritivores representing the primitive and primary mycophages with the most evolved eating habit (Hackman and Meinander 1979). In this study, the distinction of trophic levels in detail is difficult due to the approach that was given in the sporadic collection of macromycetes in good conditions and specimens that were already in early stages of decomposition, so it will be limited to establishing a statistical relationship of the presence of insects in the macromycetes of the study sites. Even so, in the state of Nuevo León there are no studies that associate insects with fungi, so the study is a good approximation that will establish a path to future research. In most fleshy fungi there are many larvae, which make it difficult to identify due to the state in which they occur. However, it can be assumed that the presence of these is due to the role played by fungi as attractants to other individuals for the dispersion of their spores. Many of the physical characteristics of fungal species may be related to the ecological niche they meet, many of these characteristics and adaptations focused on spore dispersal (Harrington and Rizzo 1999). Complying with the necessary characteristics for the development of the larvae, the insects come to deposit eggs in the fruit, they develop and begin to feed on it while they can later become adults and carry with them the spores of the fungus. Evidence of the above is the finding of Hackman and Meinander (1979) in which they analyzed *Leccinum* sporocarps that contained enough resources for the establishment of several hundred *Pegoyma* eggs, however this insect only deposited a single egg per fruit, the behavior of the insect is not with the aim of avoiding the competition of the agglomerates of larvae, but the distribution of risk by laying eggs in as many sporocarps as possible. Unlike Site 2, a greater number of organisms were found within the macromycetes located in the school Forest possibly because the appearance of these is linked to fleshy fruits that can provide greater resources to the individuals that inhabited them. Among the macromycetes that were collected with organisms within them are species of the genus such as Amanita, which had larvae within several of the specimens collected, specifically we could talk about Amanita verna which contains amanitin, which is one of the toxins of the group of the most lethal amatoxins. This type of metabolites developed by the fungus are harmful or toxic to insects in general, which would suggest that the larvae could not survive in the presence of such a compound, however evidence of tolerance to this toxin has been proven by Jaenike (1985) where he checked the tolerance of 6 species of *Drosophila* sp., 3 of the non-mycophagous species did not survive even a low concentration of the chemical, while the other 3 mycophagous species were at most barely affected by the compound. Although the presence of larvae (probably from Drosophila) is more frequent in sporadic fleshy fungi almost as a rule, those belonging to the genus Coleoptera tend to settle in polypore fruiting bodies that exist for long periods of time (Hanski 1987).

In the study, specimens of Coleoptera were found in fungal species of sporadic fruiting bodies such as Cantharellus cibarius, *Lactarius indigo* and *Lepista nuda*, which can rule out the exclusive presence of these organisms in polyporals, where the fruit lasts a long time in good condition. It is worth mentioning that these species of fungi are edible, and their ecology is limited to the presence of vegetation with which they can form mycorrhizae, so their finding in these is probably opportunistic, taking advantage of the resources available to them. Determination of abiotic variables that have a greater relationship with the presence of macromycetes in the oak and pine forest. The measurement of abiotic variables allows to identify conditioning factors in which fungi perform fruiting, the measurement of 2 years on the climate is not enough to establish these conditions or spectrum that allows fruiting, however the variables that probably had the greatest effect on macromycetes can be deferred. During the year 2016 the precipitation peaks occurred during the month of March with an accumulated rainfall in the month of 450 mm unlike the year 2017, where the month with the highest rainfall was that of September with only a little more than 50 mm. The year of 2017 was a particularly dry year with rainfall not exceeding 20 mm in the months of June, August, and October, in the other months except for September there are no records of precipitation. This contrasts with 2016, where there was precipitation in all months except February, July, and December, all of which are above 20 mm of precipitation. Along with the rains witnessed in 2016 as can be seen in Fig. 3, the minimum, average and maximum temperatures present an increasing trend as the winter months pass at the beginning of the year and it is not until the month of July that the temperature averages decrease in a stable way except for the minimum temperature in the month of December, where temperatures reached below zero degrees. In Fig. 4 of the year 2017 you can see drastic changes in temperatures, especially in the minimum temperature where in the month of March temperatures were recorded that were around 10 degrees, the curve of minimum temperatures does not follow a behavior very attached to the average and maximum temperatures, this compared to the temperatures of the year 2016. Another interesting environmental factor to analyze is the amount of solar radiation that was perceived in the years 2016 and 2017, where the year 2017 presented a peak of radiation in the month of March with 495 w/m^2 and after this the values plummet until reaching the readings of the month of October with 195 w/m^2. In contrast, 2016 showed a more constant global radiation throughout the year with the highest value in the month of June (395 w/m^2) and the minimum value recorded in the month of November with a value slightly higher than 195 w/m^2. The relationship with the

Fig. 3 Breeding chamber for insects

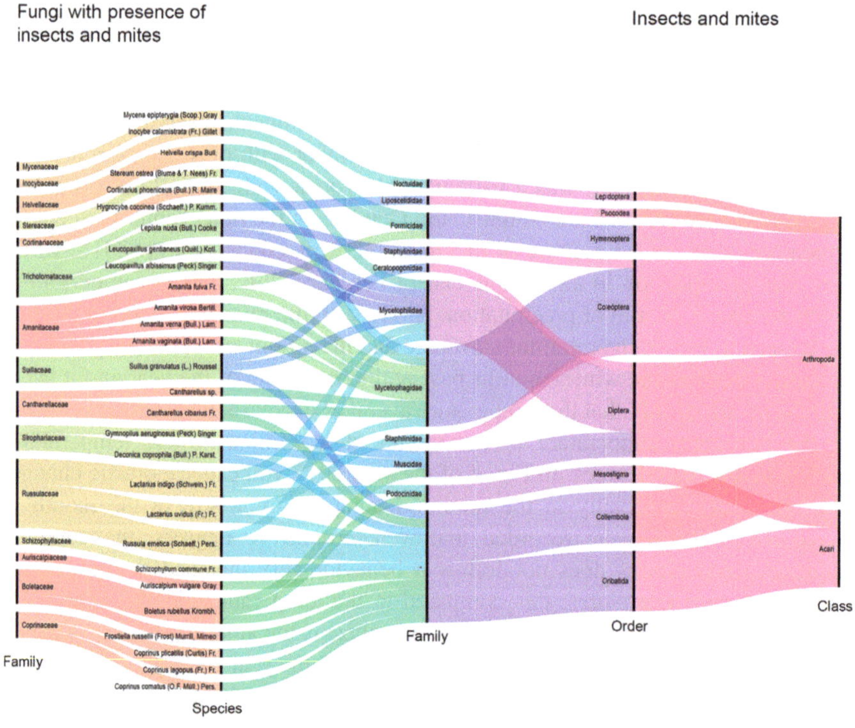

Fig. 4 Alluvial diagram of the presence of insects and mites in sporocarps. The thickness of the lines is proportional to the number of coexisting pairs. Another important observation is the relationship with the edibility of the macromycetes and arthropods found

rainfall presented in 2016 the percentage of relative humidity was higher throughout the year in contrast to 2017. The months with the highest percentage of humidity for both years were May and October highlighting the month of October 2017 with an average relative humidity of 75%, explained by being the month immediately close to the rainfall of that year, however, the amount of precipitation was very low compared to 2016.

The year 2017 presented more perceptible fluctuations, starting from the month of February as the month with the lowest relative humidity (just over 25%) until reaching October with the highest percentage. In contrast, in 2016 no values lower than 65% humidity were observed throughout the year, this can be explained with the large amount of precipitation that manifested itself during it. It was considered to integrate the evapotranspiration curves within the study because this indicator represents the amount of water in the soil that returns to the atmosphere because of the evaporation and transpiration of plants. Obviously, these changes in soil moisture and vegetation transpiration will influence the appearance of macromycetes. As expected, the year 2016 presents a stability in the values having its most critical peak in the year of July with the values of the reference evapotranspiration and potential evapotranspiration, likewise the lowest values were presented in the month of December. Back with the following year 2017 the weather station recorded very drastic changes on the potential evapotranspiration and with the absence of records in the reference evapotranspiration, all this can be explained by the absence of precipitation in this year where in itself, the values are already very low not exceeding 35 mm for the potential evapotranspiration and almost 20 mm for the reference. The lack of records from the season for the year 2017 and its strong fluctuations do not seem to make the evapotranspiration data for this year reliable (Fig. 5).

3.6 Similarity of the Sites Studied

The similarity of the sites starts from the appearance of macromycetes in the study areas, so that the lack of records may have caused strong dissociations in some of the plots (case Pine Forest - Oak * Oak Forest - Pine in Site 1). These dissonances are clearly biased from the fact that climatic conditions should limit the association of macromycetes communities. It is evident that since the plots are very close to each other and share many plant individuals, the capture of the data limits a relationship in which surely many of the species that appear in the Pine-Oak Forest and Oak-Pine Forest plots share the same mycelium under the soil.Despite the above mentioned, when evaluating the communities from a region: the vegetation, the climatic conditions, and the geography of the sites we find that there is a 48% similitude between the Ecological Campus school Forest UANL and the Ejido Puerto de Reyna. The result is not surprising considering that in the Ejido Puerto de Reyna several species of exotic characteristics were found (e.g. *Cyanoboletus* case), where being in a valley the same region served as a refuge area for these species.

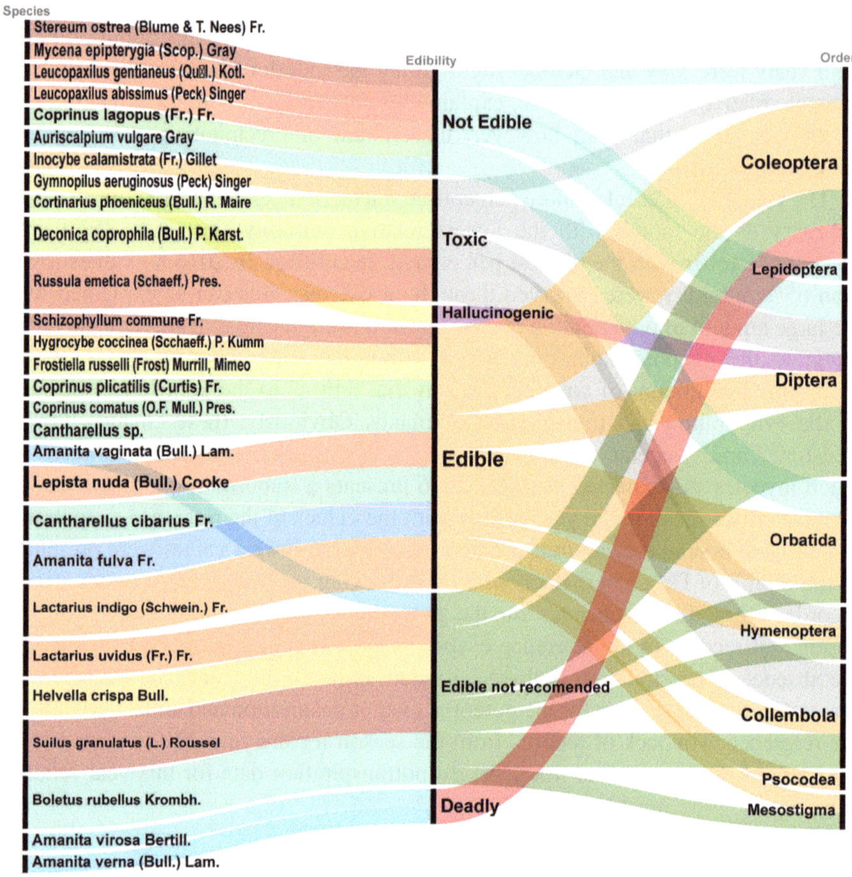

Fig. 5 Alluvial diagram of the edibility of fungi and the presence of insects

3.7 *Macromycete Species*

75 species of 58 genera and 36 families were identified, the family Tricholomataceae was the most abundant with 14 species of 9 genera followed by Agaricaceae with 5 species and 5 genera.

3.8 *Degree of Association of Insects with Sporocarps*

To identify the degree of association between the sporocarps and insects collected during the research, it was decided to group the macromycete genera with the presence or absence of insects in them. Table 6 shows the results of processing SPSS software (IBM Corp. 2016) to work the data. Later in Table 2 the chi-square statistics for the same variables are shown.

Table 2 Chi-square test for the relationship between macromycete and insect

Statistical test	Value	gl	Asymptotic significance (bilateral)	Exact significance (bilateral)
Pearson's Chi-square	107.000[a]	56	0.000	0.000
Reason for verisimilitude	147.201	56	0.000	0.000
Exact proof of Fisher	98.444			0.000
N of valid cases	107			

Table 3 Diversity of insects found by ecosystem

	Ecosystem	N	Average range
Insects	Oak Forest - Pine	16	14.28
	Pine Plantation	6	12.42
	Oak Pine Forest	4	12.00
Total		26	

Table 4 Statistical tests of chi-square by the variable of ecosystem grouping

Statistical	Insects
Chi-square	.490
gl	2
Asymptotic significance	.783

[a]Kruskal Wallis Test (Test applied)
[b]Grouping variable: Ecosystem (Test applied)

3.9 Relationship of Insects with the Type of Vegetation

Regarding the association of insects with the type of vegetation, the variables were grouped by ecosystem where Table 3 shows the count of insect appearances and Table 4 shows the statistics of the relationship by vegetation.

3.10 Identification of Species of Macromycetes and Associated Insects

The presence of 44 insects distributed in 6 orders was identified, these in turn distributed in 8 identified families. In the same sense that mites were observed within the macromycetes belonging to the orders Orbited and Messostigmado, from the latter the family Podocinidae could be identified. Table 5 shows the families of arthropods found in the collections of macromycetes.

Determination of abiotic variables that have a greater relationship with the presence of macromycetes in the oak and pine forest. Within the school forest during 2016, 43 species of mushrooms were found distributed in 17 genera; later in 2017 when repeating the collection during the same season 68 species were found within

Table 5 Families of insects and mites found in sporocarps

Class	Sub class	Order	Family
Arthropoda	Acari	Oribatidae	
		Mesostigmado	Podocinidae
	Hexapoda	Coleóptera	Staphylinidae Mycetophagidae
		Hymenoptera	Formicidae
		Lepidóptera	Noctuidae
		Collembola	NI
		Psocodea	Liposcelididae
		Diptera	Mycetophilidae Ceratopogonidae Muscidae

25 genera, of these only 14 genera appeared in both years. Table 6 presents the list of macromycetes found in the ecological campus "Bosque Escuela UANL" of Iturbide.

To improve the appreciation of the above data, the data were grouped into a sunbrust diagram in Fig. 6 where the colors differentiate the vegetation types of the mushrooms families and their species.

3.11 Analysis of the Sorensen Index

The Sorensen analysis is an indicator that allows to know the relationship of similarity between 2 variables depending on characteristics of binary variables (presence – absence) in the same way they respond to a dissimilarity. The method is based on the following fundamental principles: The sociological unit of the living being is analyzed, the unit used in the statistical treatment is a species in a population, respectively each species in each population. The similarity between two populations is expressed by the quotient of similarity,

$$IS = \frac{c}{\left(\frac{a+b}{2}\right)} * 100$$

What equals:

$$IS = \frac{2c}{a+b} * 100$$

Where:

C = Number of common species in both communities
A = Total number of species present in community A
B = Total number of species present in community B (Sørensen 1948)

For this case the results of the Sites were (Tables 7, 8, and 9) (Figs. 7 and 8):

Table 6 Insect individuals in macromycete sporocarps Sites 1 and 2 (2016–2017)

Family	Species	S 1	S2	Insects
Amanitaceae	*Amanita fulva* Fr.	X		Formicidae - Coleóptera (Mycetophagidae)
	Amanita vaginata (Bull.) Lam.	X		Larvae (Mycetopha gidae)
	Amanita verna (Bull.) Lam.	X		Larvae (Mycetopha gidae)
	Amanita virosa Bertill.		X	Larvae (Mycetopha gidae)
Auriscalpiaceae	*Auriscalpium vulgare* Gray	X		Acari (Oribatida)
Boletaceae	*Boletus rubellus* Krombh.		X	Larvae (Muscidae), Acari (Podocinid ae), Collembol a
	Frostiella russellii (Frost) Murrill, Mimeo	X		Collembola
Cantharellaceae	*Cantharellus cibarius* Fr.	X	X	Coleóptera (Mycetopha gidae) Collembola
	Cantharellus igniformis	X	X	Coleóptera (Mycetopha gidae)
Coprinaceae	*Coprinus comatus* (O.F. Müll.) Pers.	X		Acari (Oribatida)
	Coprinus lagopus (Fr.) Fr.	X		Acari (Oribatida)
	Coprinus plicatilis (Curtis) Fr.		X	Acari (Oribatida)
Cortinariaceae	*Cortinarius phoeniceus* (Bull.) R. Maire	X	X	Larvae (Mycetoph agidae) 2016
Helvellaceae	*Helvella crispa* Bull.	X		Formicidae y Larvae (Mycetophili dae)
Inocybaceae	*Inocybe calamistrata* (Fr.) Gillet	X	X	Formicidae
Mycenaceae	*Mycena epipterygia* (Scop.) Gray	X	X	Lepidoptera (Noctuidae)
Russulaceae	*Lactarius indigo* (Schwein.) Fr.	X		Larvae (Mycetophili dae), Acari (Podocinidae), Coleóptera (Mycetophagidae)
	Lactarius uvidus (Fr.) Fr.	X	X	Coleóptera (Staphilini dae), Collembol a
	Russula emetica (Schaeff.) Pers.	X	X	Larvae (Mycetophili dae), Acari (Oribatida), Collembola
Stereaceae	*Stereum ostrea* (Blume & T. Nees) Fr.	X		Coleoptera (Mycetopha gidae)
Schizophyllaceae	*Schizophyllum commune* Fr.	X		Acari (Oribatida)
Strophariaceae	*Deconica coprophila* (Bull.) P. Karst.	X		Acari (Oribatida), Larvae (Muscidae)
	Gymnopilus aeruginosus (Peck) Singer	X		Larvae (Muscidae)
Suillaceae	*Suillus granulatus* (L.) Roussel	X	X	Larvae (Ceratopogo nidae, Collembola); y Mycetophilidae)

(continued)

Table 6 (continued)

Family	Species	S 1	S2	Insects
Tricholomataceae	*Hygrocybe coccinea* (Scchaeff.) P. Kumm.	X		Psocodea (Liposcelidi dae)
	Lepista nuda (Bull.) Cooke	X	X	Coleóptera (Staphylinidae), Larvae (Mycetophili dae)
	Leucopaxillus albissimus (Peck) Singer	X		Larvae (Mycetophili dae)
	Leucopaxillus gentianeus (Quél.) Kotl.	X		Larvae (Mycetophili dae)

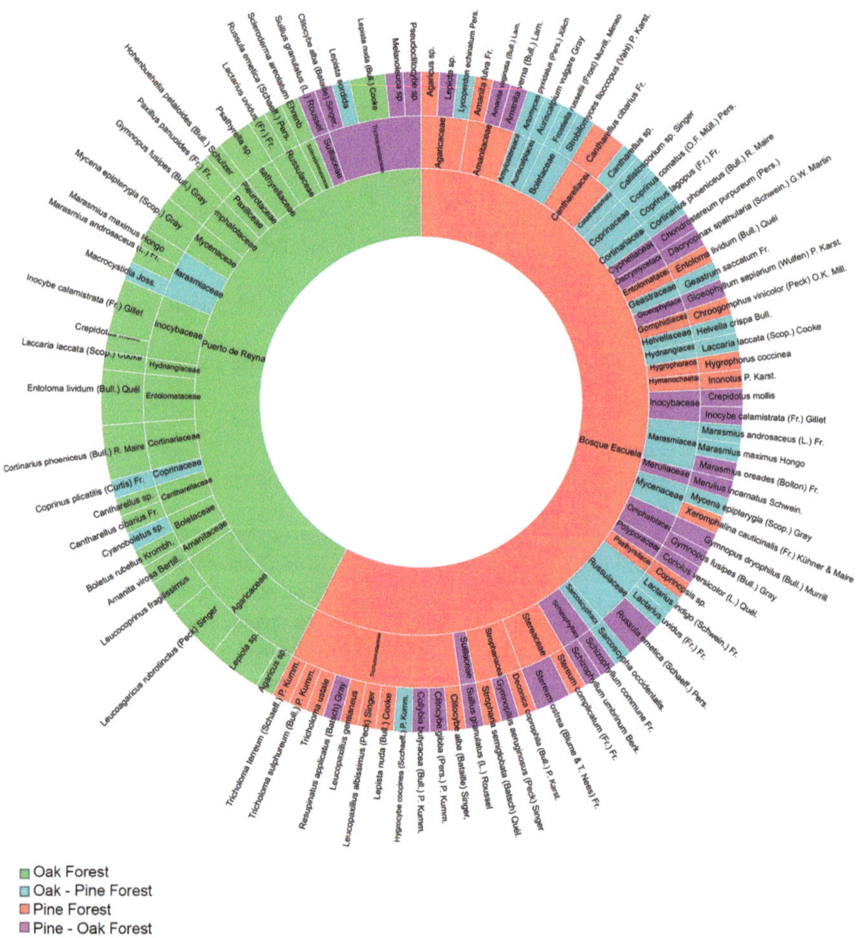

Fig. 6 Sunbrust diagram of the macromycetes found by type of vegetation and collection sites

Table 7 Similarity between plots of Site 1

Plots at forest school UANL Iturbide NL	Percentage of similarity
Oak-Pine Forest * Pine	8.00%
Pine Forest - Oak *Oak Forest - Pine	0.00%
Pine * Oak - Pine Forest	8.33%

Table 8 Similarity between plots of Site 2

Plots at Ejido Puerto de Reyna, Galeana, N.L.	Percentage of similarity
Oak Forest - Pine * Oak Forest	48%
Pine Forest - Oak * Oak Forest - Pine	48%
Encino Pine Forest * Oak Forest	36%

Table 9 Similarity between the sites studied

Similarity Bosque Escuela UANL Iturbide NL * Ejido Puerto de Reyna	**41%**

4 Conclusions

Within the study, an association between sporocarps and the presence of insects with a p-value < 0.05 is statistically verified. The appearance of the fructifications of the fungi is related to the type of vegetation where they are found, anthropocentric disturbances and climatological changes over the years. The oak forest of the Ejido Puerto de Reyna was the one with the greatest diversity of macromycetes (25) due to the high degree of dominance of Quercus spp at the site. This area has silvograzing livestock activity affecting the composition of the vegetation. In contrast, the pine plantation of the Ecological Campus "Bosque Escuela" presented the least diversity being related to the young age of the plantation. The macromycete families found were: Agaricaceae, Amanitaceae, Amylostereaceae, Auriscalpiaceae, Boletaceae, Cantharellaceae, Catathelasmataceae, Coprinaceae, Cortinariaceae, Cyphellaceae, Dacrymycetaceae, Entolomataceae, Geastraceae, Gloeophyllaceae, Gomphidiaceae, Helvellaceae, Hydnangiaceae, Hygrophoraceae, Hymenochaetaceae, Inocybaceae, Marasmiaceae, Meruliaceae, Mycenaceae, Omphalotaceae, Polyporaceae, Paxillceae, Pleurotaceae, Psathyrellaceae, Russulaceae, Sarcoscyphaceae, Stereaceae, Schizophyllaceae, Strophariaceae, Sclerodermataceae, Suillaceae, Tricholomataceae. The Tricholomateace family was the most abundant with 14 species spread over 9 genera with most of the specimens found in pine plantation and oak-pine forests. The macromycete with the greatest diversity of insects was Russula emetica presenting individuals of Larvae (Diptera), Thysanoptera, as well as individuals belonging to Acari, which despite not being insects, appeared frequently in the macromycetes. It can be said

Fig. 7 *Sporocarps of Marasmius; Merulius, Mycena Xeromphalina, Gymnopus, Coriolus, Paxillus, Hohenbuehelia, Psathyrella* and *Lactarius*

that the appearance of these organisms in macromycetes were opportunistic in nature taking advantage of the availability of the resource (Hanski 1989). The conditions of low precipitation in 2017 were manifested in the low diversity of macromycetes collection and the appearance of insects in them.

Fig. 8 Insects associated with macromycete sporocarps: specimens of the families Liposcelididae, Collembola, Formicidae, Noctuidae, Staphylinidae an Mycetophagidae

References

Annila E (1982) Diapause and population fluctuations in *Megastigmus specularis* Walley and *Megastigmus spermotrophus* Wachtl. (Hymenoptera, Torymidae). Ann Entomol Fenn 48(2): 33–36

Bässler C, Müller J, Dziock F, Brandl R (2010) Effects of resource availability and climate on the diversity of wood-decaying fungi. J Ecol 98(4):822–832. https://doi.org/10.1111/j.1365-2745.2010.01669.x

Beaver RA (1989) Insect-fungus relationships in the bark and ambrosia beetles. Insect- Fungus Interact 121:143

Bruns TD (1984) Insect mycophagy in the Boletales: fungivore diversity and the mushroom habitat. In: Wheeler Q, Blackwell M (eds) Fungus-insect relationships, perspectives in ecology and evolution. Columbia University Press, New York, pp 91–129

Castillo-Guevara C, Lara C, Pérez G (2012) Micofagia por roedores en un bosque templado del centro de México. Rev Mex Biodivers 83(3):772–777. https://doi.org/10.7550/rmb.27445

Cerda LJ, Villarroel Del PL (2007) Interpretación del test de Chi-cuadrada (X 2) en investigación pediátrica. Rev Chil Pediatr 78(4):414–417. https://doi.org/10.4067/s0370-41062007000400010

Challenger A, Soberón J (2008) Los ecosistemas terrestres. Capital Natural de México, Vol. I: Conocimiento Actual de La Biodiversidad, I, 87–108.

Dighton J (2003) Fungi in ecosystem processes. https://doi.org/10.1201/9780203911440

Fergus CL, Fergus C (2003) Common edible and poisonous mushrooms of the northeast. www.stackpolebooks.com

Flores Cavada E, Carrillo Parra A, Wehenkel CA, Garza Ocañas F, Hernández Díaz JC (2018) Diversidad de macromicetos en bosques de pino en el municipio Madera, Chihuahua. Rev Mex Cienc For 9(50). https://doi.org/10.29298/rmcf.v9i50.240

Gadd GM, Watkinson SC, Dyer P (2007). Fungi in the environment

Gaitán JJ, López CR, Bran DE (2009) En La Estepa Patagónica 27(2):261–270

Garza F (1986) Hongos Ectomicorricicos en el Estado de Nuevo León. Rev Mex Micol 2:197–205

Hackman W, Meinander M (1979) Diptera feeding as larvae on macrofungi in Finland. Ann Zool Fenn 16(1):50–83

Hanlin RT (2004) Mushrooms of West Virginia and the central Appalachians (review). Nat Plants J. https://doi.org/10.1353/npj.2005.0006

Hanski I (1987) Nutritional ecology of dung- and carrion-feeding insects. In Nutritional ecology of insects, mites and spiders (pp 837–884)

Hanski I (1989) Fungivory: fungi, insects and ecology. In: Insect-fungus interactions. The Royal Entomological Society of London, London. https://doi.org/10.1016/B978-0-12-751800-8.50008-2

Harley JL (1989) The significance of mycorrhiza. Mycol Res 92(2):129–139. https://doi.org/10.1016/S0953-7562(89)80001-2

Harrington TJ, Rizzo DM (1999) Defining species in the fungi. In: Structure and dynamics of fungal populations. Springer, Dordrecht, p 348. https://doi.org/10.1007/978-94-011-4423-0_3

Hongos de la Tierra Declaración de Córdoba, 6 (2007)

IBM Corp (2016) IBM SPSS Statistics for Windows, Version 24.0. In 2016. Ihering, H. von. (1898). Die Anlage neuer Colonien und Pilzgarten bei Atta sexdens. Zoologischer Anzeiger, 238–245

Index Fungorum (n.d.) http://www.indexfungorum.org/Names/Names.asp

INEGI (2005) Guía para la Interpretación de Cartografía Climatológica. Guía Para La Interpretación de Cartografía Climatológica:21–33

INEGI (2009) Prontuario de Información Geográfica Municipal.

Jaenike J (1985) Parasite pressure and the evolution of Amanitin tolerance in drosophila. Evolution 39(6):1295–1301. https://doi.org/10.2307/2408786

Kauserud H, Heegaard E, Semenov MA, Boddy L, Halvorsen R, Stige LC, Sparks TH, Gange AC, Stenseth NC (2010) Climate change and spring- fruiting fungi. Proc R Soc B: Biol Sci 277(1685):1169–1177. https://doi.org/10.1098/rspb.2009.1537

Köppen W (1918) Klassifikation der Klimate nach Temperatur, Niederschlag und Jahresablauf. Petermanns Geographische Mitteilungen 64:193–203

Marks GC (2012) Ectomycorrhizae: their ecology and physiology. Elsevier Science. https://books.google.com.mx/books?id=SVPpx40FS6QC

Márquez Luna J (2005) Técnicas de colecta y preservación de insectos. Boletín Sociedad Entomológica Aragonesa 37:385–408

Maser C, Claridge AW, Trappe JM, Krebs CJ (2008) Trees, truffles, and beasts: how forests function. In Trees, truffles, and beasts: how forests function. https://www.scopus.com/inward/record.uri?eid=2-s2.0-84906443805&partnerID=40&md5=e8642af9191c1586a18501410f11c2ab

Maser C, Maser Z, Molina R (1988) Small-mammal mycophagy in rangelands of central and southeastern Oregon. J Range Manag 41(4):309–312. http://www.jstor.org/stable/10.2307/3899385

Mycobank (n.d.). http://www.mycobank.org/

Ostry NA, O'Brien JG, Anderson ME (2010) Field guide to common macrofungi in eastern forests and their ecosystem functions. USDA Forest Service Northern Research Station General Technidal Report, NRS-79, 1–90. https://doi.org/10.2737/NRS-GTR-79

Pereira H, José M (2015) Potencial forrajero de la vegetación del bosque húmedo tropical como un sistema silvopastoril en laregión de barlovento del estado miranda, vol 1887. Universidad Central de Venezuela

Quiñónez-Martínez M, Lebgue Keleng T, Lavín-Murcio P, Bernal-Carrillo S (2013) Influencia del disturbioen la riqueza de hongos ectomicorrizógenos en los bosques de Chihuahua. Ciencia En La Fronetera, XI, pp 9–16

Region USFSE (1980) Root characteristics of some important trees of eastern forests: a summary of the literature. Department of Agriculture, Eastern Region, Forest Service, U.S. https://books.google.com.mx/books?id=oLiRG-38UN8C

Repetto Giavelli F (2012) Análisis de la regeneración natural de la vegetación en taludes de caminos de la Ruta Y-85, Parque Karukinka, Tierra del Fuego Chile. Anales Instituto Patagonia (Chile) 40(2):55–65. https://doi.org/10.4067/S0718-686X2012000200005

Rodrigo, J. A. (2016) Kruskal-Wallis test. https://rpubs.com/Joaquin_AR/219504

Singer R, Guzman G (1978) Identificacion de los hongos comestibles, venenosos y alucinantes. Mycologia. https://doi.org/10.2307/3759048

Smith SE, Read DJ, Kiers ET, Duhamel M, Beesetty Y, Mensah JA, Franken O, Verbruggen E, Fellbaum CR, Kowalchuk GA, Hart MM, Bago A, Palmer TM, West SA, Vandenkoornhuyse P, Jansa J, Bücking H (2008) Mycorrhizal symbiosis. Soil Sci Soc Am J 137. https://doi.org/10.1097/00010694-198403000-00011

Sørensen T (1948) A method of establishing groups of equal amplitude in plant sociology based on similarity. Kong Dansk Vidensk Selsk Biol Skr 5(4):1–34

Stamets P (2005) Mycelium running how mushrooms can help save the world. In Mycelium running: how mushrooms can help save the world

Sturgeon WE (2018) Appalachian mushrooms: a field guide. Ohio University Press. https://books.google.com.mx/books?id=NHhyDwAAQBAJ

Vargas LB (1999) Caracterización de la productividad y estructura de Pinus hartwegii Lindl en tres gradientes altitudinales en el cerro Potosí, Galeana, Nuevo León

Epilogue

The readers will have become familiar with the importance of sustainable management of forest resources, their diversity, taxonomic identification, ecology, and sociology. This book arises from the authors concern to contribute and publish the research that each one carries out to contribute to scientific knowledge and education for students and the public in general. Ecosystems and their species play a very important role for human health and their conservation and rational and sustainable use is necessary for the survival of all. The sustainable use and management of species is part of a non-destructive alternative to value natural resources. The social educational process contributes positively to the multifunctional management of ecosystems and has a positive impact. Species in the forest ecosystem and soil, freshwater aquaculture systems and grasslands play an important role for their own balance and that of the man who takes advantage of them. As it is well known, there are other factors that contribute to complexity and cause of ecosystems deterioration. Climate change, long periods of drought, the frequency and size of forest fires, increasing strong frosts in winter as well as tropical storms, tornadoes and hurricanes are already part of the annual global landscape and affect species, the global economy, water availability and propitiate human health issues. Thus, this book modestly aims to show some of the investigations that the authors have focused on their respective specialty areas with an educational academic purpose for the knowledge and improvement of the sustainable use of natural resources and the conservation of nature.

The editor

F. Garza-Ocañas (ed.), *Sustainable Management of Natural Resources*, Earth and Environmental Sciences Library, https://doi.org/10.1007/978-3-031-33394-1

MIX
Papier aus verantwortungsvollen Quellen
Paper from responsible sources
FSC® C105338

If you have any concerns about our products,
you can contact us on
ProductSafety@springernature.com

In case Publisher is established outside the EU,
the EU authorized representative is:
Springer Nature Customer Service Center GmbH
Europaplatz 3, 69115 Heidelberg, Germany

Printed by Libri Plureos GmbH
in Hamburg, Germany